A Farmer's Guide to Climate Disruption

A FARMER'S GUIDE TO CLIMATE DISRUPTION

Rebekah L. Fraser

Bee Books
New Haven

A Farmer's Guide to Climate Disruption
Copyright © 2018 by Rebekah L. Fraser
All rights reserved. Printed in the USA.
For information, visit BeeBooks.org
Cover photo by Rebekah L. Fraser

The Library of Congress Cataloging-in-Publication Data
is available upon request.

ISBN 978-1-7326187-4-9 (hardcover)
ISBN 978-1-7326187-2-5 (paperback)
ISBN 978-1-7326187-3-2 (e-book)

First edition: December 2018

ADVANCE PRAISE

"A Farmer's Guide to Climate Disruption weaves together science, anecdotes, and practical tips to explain the many issues farmers face as they try to continue growing the food we need in spite of water scarcity, more violent weather events, ecological change and sea level rise. Farmers and others concerned about agriculture will find this book a valuable and accessible resource in confronting the many challenges stemming from global climate disruption."

~ David M. Driesen, Professor, Syracuse University

"Very user-friendly in its organization and well-written. We are all seeing the challenges of weather extremes and how to adapt agricultural practices to cope with these changes and develop resilient landscapes. A Farmer's Guide to Climate Disruption brings many resources and ideas into one cohesive whole."

~ Ann J. Adams, PhD, Executive Director, Holistic Management International

"An incredibly important book. We are already being affected by a changing climate and the role of farming will be essential to our survival. A Farmer's Guide to Climate Disruption is a very valuable resource for farmers, ranchers, and land managers as it offers a multitude of tools, techniques and evidence-based practices that can be used to build resilience and adaptation."

~ Andre Leu, International Director, Regeneration International

"Rebekah is a much-needed, informative voice in the subject of climate change. Her expertise shines a light on the severity of the issues at hand. She's able to effectively communicate those concerns that affect everyday growers."

~Michael Freeze, former editor Growing Magazine

"A pleasure to read... We need to have more discussion about what we are doing and where we are going in agriculture to meet food needs and environmental goals of clean water and air."

~ Jerry Hatfield, , USDA, co-author 3rd National Climate Assessment

This book is dedicated to the many farmers and home gardeners who taught me what it means to tend the earth with care, and to my daughter, for whom I strive to build a healthier world.

CONTENTS:

Foreword by Andre Leu   i
Introduction: Why this book? Why now?   iv

PART I. FOOD SECURITY AT YOUR FINGERTIPS

1. Weathering Climate Disruption   3
2. Creating Food Security in Our Changing Climate   6

PART II. COMMUNICATING CLIMATE CHANGE

3. Speaking in Tongues: The Language of Climate Disruption   17
4. To Teach or Not to Teach: Educating Consumers   21

PART III. DISASTER AT YOUR DOOR

5. How Proactive Farmers Are Resilient in Extreme Weather   26
6. Now What?   30
7. Prepare for the Worst: Tips for Recovering   38

PART IV. SOIL: THE CLIMATE BELOW YOUR FEET

8. Profile of a Misunderstood Substrate   49
9. Soil Solutions: Cover Crops   53
10. No-Till for Soil Health?   56

PART V. BEYOND CARBON

11. Nitrogen Must Be Managed for Climate Resilience   65
12. New Tech in N^2O Sampling   68

PART VI. WATER

13. Adapting When There's Too Much or Not Enough   73
14. The Art of Water: New Concepts in Irrigation   79

PART VII. UNDER PRESSURE: PESTS, PATHOGENS & WEEDS

15. Pests & Pathogens Are Adapting: What This Means   87
16. How to Adapt to Adaptive Pests and Pathogens   91
17. Learning from Weeds   95

PART VIII. CROPS: WHAT HORTICULTURISTS KNOW

18. Preparing Slow-Growth Crops for a Changing Climate   101
19. How California Almonds Are Thriving   105
20. Breeding for the New Climate   109

PART IX. CLIMATE SMART FARMING METHODS

21. Pollination Challenges and Strategies   114
22. Regenerative Agriculture   117
23. Push Pull: How Peas & Grasses Fight Climate Disruption   121
24. The Post-Carbon City and Farming   125
25. The Post-Carbon Farming System in Practice   130

RESOURCES   140
A NOTE FROM THE AUTHOR   142
ACKNOWLEDGEMENTS   143
BIBLIOGRAPHY   147
ABOUT THE AUTHOR   154

FOREWORD

By Andre Leu, International Director, Regeneration International

Rebekah has written an incredibly important book. We are already being affected by a changing climate and the role of farming will be essential to our survival. We need to ensure that farming can adapt and be resilient to the increase in extreme weather events so that there is enough food to feed all of us. This book is a very valuable resource for farmers, ranchers and land managers as it offers a multitude of tools, techniques and evidence-based practices that can be used to build resilience and adaptation.

Many people are not concerned if the world becomes a few degrees hotter, however they do not understand that average temperatures are not the main problems. The climate extremes are the existential problems that agriculture faces. The world is already one degree C ($1.8\,^{\circ}F$) warmer than 100 years ago. It takes the energy equivalent of billions of nuclear bombs to heat up the world by $1\,^{\circ}C$ ($1.8\,^{\circ}F$). I am using this violent metaphor so people can understand that all this extra energy is violently pushing our weather to more extremes. Storms are getting stronger; colder in winter and hotter and wetter in summer. Droughts are more frequent and longer. When the rain comes, it is often heavy and destructive and out of season. Climate disruption is a better term to describe what we are already experiencing. This is leading to more crop failures. As farmers we need to cope with this if we and humanity are to survive; after all everybody has to eat.

The atmospheric CO_2 level had been increasing at 2 parts per million (ppm) per year. The level of CO_2 reached a new record of 400 parts ppm in May 2016. This is the highest level of CO_2 in the atmosphere for 800,000 years. However, in 2016, despite all the commitments countries made in Paris in December 2015, the levels of CO_2 entering the atmosphere increased to 3.3 ppm creating a new record.

According to the World Meteorological Organization, "Geological records show that the current levels of CO_2 correspond to an "equilibrium" climate last observed in the mid-Pliocene (3–5 million years ago), a climate that was 2 to 3 °C (5.4 to 9 °F) warmer, where the Greenland and West Antarctic ice sheets melted and even some of the East Antarctic ice was lost, leading to sea levels that were 10–20 m [30 to 65 Ft] higher than those today."

Global sea levels rises will cause the atoll island countries, large parts of Bangladesh, Netherlands, coastal USA, New York, New Orleans, Miami, London, Manila, Bangkok, Jakarta, Shanghai, Singapore, and other low lying areas to go under water, causing a huge refugee crisis for over a billion people.

Even if the world transitioned to 100 percent renewable energy tomorrow, this will not stop the temperature and sea level rises because it will take more than 100 years for the CO_2 levels to drop.

The fact is we have speed up the transition to renewable energy and we have to make a great effort to drawdown the CO_2 in the atmosphere. Agriculture has a very important role, in doing this as it can drawdown enough CO_2 and store it in the soil to reverse climate disruption.

Soils are the greatest carbon sink after the oceans. According to Professor Rattan Lal, there are over 2,700 gigatons (Gt) of carbon stored in soils. Soils hold more than double the amount of carbon than is stored in the atmosphere (848 Gt) and forests (575 Gt) combined. There is already an excess of carbon in the oceans that is starting cause a range of problems. We cannot put any more CO_2 in the atmosphere or the oceans. Soils are the logical sink for carbon.

Most agricultural systems lose soil carbon with estimates that agricultural soils have lost 50 percent to 70 percent of their original soil organic carbon (SOC) pool, and the depletion is exacerbated by further soil degradation and desertification. Agricultural systems that recycle organic matter and use crop rotations can increase the levels of SOC. This is achieved through techniques such as longer rotations, cover crops, green manures, legumes, compost, organic mulches, biochar, perennials, agro forestry, agroecological biodiversity and livestock on pasture using holistic grazing systems. These systems are starting to come under the heading of Regenerative Agriculture because they regenerate SOC.

We have enough evidence-based science and practice to show the widespread adoption of these regenerative methods to increase SOC can drawdown enough CO^2 to reduce its levels every year and bring the climate back to preindustrial revolution temperature. The widespread scaling up of these farming methods could achieve this by 2050 so that world will avoid catastrophic climate disruption.

The immediate goal must be to stabilize the CO^2 in the atmosphere at 400 ppm to prevent any further increases in the extreme weather events caused by climate disruption. Ideally, this should be done by capping the current emissions and adopting a combination of renewable energy and energy efficiency. However, under the Paris agreement this will not happen until 2030 at the earliest.

Regenerative Agriculture can change agriculture from being a major contributor to climate disruption to becoming a major solution. The widespread adoption of these systems should be made the highest priority by governments, international organizations, industry, farming and climate change organizations.

INTRODUCTION
WHY THIS BOOK? WHY NOW?

When I first heard about global warming in the 1990s, I didn't think much of it. Although I considered myself an environmentalist at the time, and I was an ardent supporter of organic agriculture, I didn't see the threat as real. A New England native, I had moved to Southern California, where I worked in a climate-controlled office at a manufacturing company. Even as I was enjoying my beach time, I felt generally disconnected from nature.

It took moving to rural Massachusetts to heighten my senses. Taking daily walks in the woods, I began to observe the subtle changes in my environment as they were happening. Entering journalism, my environmental awareness expanded. An assignment to cover electronic waste for *EM*, the Air & Waste Management Association's magazine, opened my eyes to a variety of waste management issues — including what to do with food waste. I became fascinated by composting and started a blog about creative waste diversion, which led to an assignment from *Vegetarian Times* about composting for urbanites. In the fall of 2008, I was hired by *Growing* Magazine to write a feature about the art and science of seeds. Because the article was so well received, the editor invited me to write a monthly column about seed science. As a film/screenwriting scholar, I was in strange territory, exploring the worlds of genetic engineering and traditional seed breeding.

In the five years that I wrote the seed research column, I also wrote several feature articles for Growing and its sister publication, *Farming: The journal of Northeast Agriculture*. I visited farms and land-grant universities. I connected with agriculture professionals from all over the world. I noticed a trend: researchers were expressing concern about climate change (which many now call climate disruption).

They weren't spouting the message of Bill McKibben or 350.org. They weren't quoting Al Gore. The researchers I met were observing the subtle changes in their fields, and in their crops, both at home and work, as the overall warming of the globe was causing ever-more erratic changes in the worldwide climate. Researchers were identifying genes that cause various responses to temperature, moisture or other climatic shifts. They were seeing alterations in pest behavior and proliferation. All these observations led the researchers I met to ask: how will climate disruption affect food security, not just in my area, but region-to-region, country-to-country, across the globe? They were concerned about all aspects of food security:

- Production & Yield
- Pest & Disease Pressure
- Food Quality
- Population

Because I am obsessed with food (specifically fruits, nuts, and vegetables), and I am always concerned that everyone has enough to eat, I became obsessed with the connection between agriculture and climate disruption.

According to current projections, global population will be 30 percent higher in 2050. Because of this and anticipated dietary changes, growers will need to produce 50 percent more food annually by 2050. Farmers will need to increase productivity without expanding the land base (there is little land base on which to expand) or causing further environmental degradation. Expanding food production by 50 percent in 30 years in a changing climate requires an "all hands on deck" approach. But what can growers do to shape the future of our global food supply? What happens if population growth exceeds expectations?

Those questions led me to pen *Growing's* first-ever monthly column about climate disruption, "Changing Climate," which ran in each issue of the publication from January 2016 until the magazine folded in December 2017.

In preparing for the column, I interviewed some of the world's top researchers in the field of agriculture. I could not get those same interviews today because the U.S.' executive office banned all Department of Agriculture (USDA) scientists from discussing climate change with journalists in 2017. When *Growing* and *Farming* folded at the end of 2017, I took comfort in knowing the archives would be available to farmers in perpetuity, because this information is imperative for food growers who want to adapt and thrive in the changing climate. I was wrong. The publisher took the archives offline. To make this information available to you, I have compiled all of my climate columns here, along with some relevant feature articles. Where needed, I have updated the pieces. I'm also adding some previously unpublished material, such as the highly informative and inspiring interviews I conducted with Jack Algiere of Stone Barns Agricultural Center that I intended to publish in *Growing* in January 2018 (I interviewed Algiere before I learned the magazine would fold.)

May this information bless your crops and help you feed your loved ones and your community for the next seven generations.

PART I

FOOD SECURITY AT YOUR FINGERTIPS

"From climate change to poverty, hunger or health: agriculture plays a major role in shaping our future."

~ Hermann Lotze-Campen, chair of PIK research domain Climate Impacts and Vulnerabilities

1.
WEATHERING CLIMATE DISRUPTION
Shaping the Future

Warmer winters, superstorms, floods, droughts, new pests and diseases… If you've experienced any of these issues on your farm in the last ten years, you may have thought it was an isolated incident, a hazard of farming in your particular environment on your specific piece of land. That may be true. However, if you connect with other growers in other areas around the country and the globe, you will probably hear similar stories. This web of "isolated incidents" has been happening so much more frequently in the last several years, and with such greater intensity, that scientists feel confident saying the incidents are connected, and are the result of a single global event: the overall rise in temperature. We now know this as "Climate Disruption."

A word about scientists…
   Scientists are cautious by nature. If you read scientific literature, you find words like, "Suggests," and "Likely," and "Probable." Scientists are hesitant to make big pronouncements about anything. Instead, they will research more, explore more and wait for more evidence to pile up before making a definitive statement. Case in point: it took scientists decades to announce the trend of global warming. Did this happen in the last few years? No. It happened in 1988, when NASA climate scientist Dr. James Hansen testified

to the US Congress that the earth's atmosphere was warming. He wasn't postulating. He was reporting proven scientific fact. In the quarter century since that meeting, Hansen's warnings came true. Yet people are still debating whether climate disruption is real. It took decades of mounting evidence for researchers to say, "Yes. This doesn't seem to be happening. This is happening."

What does this mean for you?
 I like to think of climate disruption as an illness. There are symptoms (deluges, droughts, etc…), and there is the root cause (increased carbon released into the atmosphere.) As with any illness, if you focus only on treating the symptom, you may find temporary relief, but you are unlikely to cure the disease. You and I need to address the cause. Based on where you are and where I am, and your profession (farmer) and my profession (writer), we will utilize different treatments. However, we're all faced with the same disease, and we all have the power to treat the cause.

What if we don't act?
 According to a study by the Potsdam Institute for Climate Impact Research, global food demand will double by mid-century; in particular the demand for animal products will rise rapidly. Contrary to what you may be thinking about supply and demand, this is not good news, as animal products require even more land and crop resources. I have interviewed numerous agricultural researchers on the topic of climate disruption and farming. While some are optimistic about the possibilities for growers and ag researchers to find the right mix of treatments, prophylactics, and creative solutions, they all said the same thing: in just over 30 years, farmers will need to feed a global population that is expected to grow from 7.6 billion (our current population) to at least 9.5 billion. In a healthy climate, this wouldn't be so daunting. However, we are not living in a healthy climate. We have issues with water pollution, air pollution and soil degradation. Scientists report expanding ranges of insects, diseases, and weeds. If we make no changes, extreme weather events are predicted to increase dramatically. You know better than I

that if you lose your crop to an unexpected hail storm, it's hard to feed the people you normally feed, never mind to produce food for even more people.

How can I help?

I'm a writer, not a scientist, nor a farmer. While you have some heavy lifting ahead of you, I hope this book will support you with the information you need to thrive during what promises to be some very challenging times.

In these pages, I address a variety of climate disruption related topics, and explore how they are related to growing food. Through this book, you'll meet climatologists, soil scientists, experienced farmers, geneticists, and other researchers and practitioners. These are people working to find the most effective ways to get at the root cause of climate disruption, without forcing you to sacrifice your way of life completely. As USDA researcher Jerry Hatfield says, "It's not just an esoteric exercise we're working on in the science community. We really do have the producer in mind to help them grow crops effectively."

We'll learn why some agricultural researchers are eating the words of their forebears, as they realize the information they delivered to farmers in the last century actually contributed to climate disruption. We'll learn about the importance of biodiversity and what that means in an agricultural setting. We'll learn about changing growing regions, irrigation strategies, soil health, rising sea levels, new cultivars that may thrive in your changing climate, techniques for growing in unpredictable weather, and how to use your influence at the local, state and national level to ensure the policies enacted are the most sound and logical under the circumstances. Perhaps most important, we'll learn how shifting your farming practices can actually combat global warming.

Weathering climate disruption is something we can only do together.

2.

CREATING FOOD SECURITY IN OUR DISRUPTED CLIMATE

"Agriculture has a tremendous challenge. It's not something we can wait until 2025 to start thinking about. We need to start right now."

~Jerry Hatfield, USDA-ARS, 2014

Emerging data reveals plants are more sensitive than previously thought. An average temperature rise of just six degrees Fahrenheit can severely affect quantity and quality of yields. As temperatures rise, many researchers in the ag community feel an urgency to bring the tools from their particular area of expertise to a newfangled barn-raising. Of course, rather than putting up a building in a day, the challenge is to collaborate for at least the next three decades to put up food for a growing population.

"If we want to feed the world, in the next thirty-five years, we have to produce as much food in that period of time as we produced in the last 1500 years," said Hatfield, an agricultural climatologist with USDA-ARS who has spent over forty years researching the effects of climate on crop growth. "Agricultural production is going to have to rapidly increase to meet that challenge, with more extremes in weather."

The concern about climate disruption is expanding rapidly as our weather becomes more extreme: too cold, too hot, too dry, and too wet, even within the same growing season. Each of these extreme weather situations brings additional challenges, as well.

Aspects of Climate Disruption as Related to Agriculture

Climate disruption is a puzzle, not a single problem with a single clear-cut solution. The pieces of the puzzle include:

Global warming: as overall temperatures rise, productivity decreases. Plants have a minimum temperature, an optimum temperature at which they grow best, and a maximum temperature, above which they die. The negative slope of the curve between optimum and maximum temperature is steep. Plants growing above their optimum are generally stunted and yields are small. Although some crops are starting to grow in previous cooler areas due to warming, yields decrease when plants must survive in the upper range of their temperature tolerance level.

Extreme weather: colder winters and hotter summers with more intense, less frequent precipitation.

Unpredictable, yet more frequent extreme weather events – snow storms in the Southeast US, tornadoes in the Northeast, Hurricanes: even having the ability to predict extreme temperature and rainfall events would benefit growers, by allowing them to adjust planting dates, to protect the pollination phase.

Pollination phase interruption: The pollination phase is a critical stage in crop growth. As the plant produces pollen or begins to set fruit, exposure to a high temperature event can destroy the pollen, and basically render the plant sterile. The end result is zero crop production.

Water supply: plants are more susceptible to high temperature stresses if they are either inadequately watered or over-watered.

"One of the things that we have to do is start thinking about adaptation strategies and realizing what differences are within species for some of these responses," said Hatfield. For decades, Hatfield's focus has been to find mechanisms growers can employ to help offset various climate stresses. For example, changing varieties based on when they pollinate during the day has a tremendous impact, and one doesn't have to disruption anything other than the variety one plants.

The researchers interviewed for this piece agreed: the nature of this problem is not something agriculture can use traditional approaches to solve. Multiple parties from multiple sectors need to come together at the onset, come up with solutions, and anticipate how they'll be applied. "Otherwise, we're going to end up with lots of different food shortages in lots of places," warns Hatfield.

Approach: Starting Small

"The time it takes plants to adapt naturally, or through evolution is much longer than the timeframe we're looking at," says Kent Bradford, Distinguished Professor of Plant Sciences and Director of the Seed Biotechnology Center at UC Davis. Bradford's recent research focuses on heat tolerance of lettuce seed germination.

In 2013, Bradford identified NCED4, a gene involved in inhibition of germination of lettuce seed. In most lettuce seed, hydration at high temperatures activates NCED4, which then creates an enzyme involved in synthesizing the hormone abscisic acid, which ultimately inhibits germination. However, the particular gene from a wild lettuce line, accession UC96US23, does not respond to high temperature by activating NCED4, and so doesn't make the hormone. Without the presence of abscisic acid, there is nothing to inhibit germination. Therefore, UC96US23 seeds germinate at very high temperature. This ability to germinate at high temperatures is rare among the L. serriola species.

"We're interested in understanding how temperature affects the expression of this gene, and what mechanisms plants use to sense temperature, because many growing regions are experiencing higher temperatures with climate change," Bradford explains.

One of a few researchers focusing on preparing leafy green vegetables for the extreme weather shifts expected in the next forty years, Bradford realizes no matter how sturdy it becomes, lettuce will never be a world-feeder. Still, thanks to this research, scientists now know there is a specific mechanism that senses temperature and responds by expressing NCED4. Currently, Bradford and his team are trying to determine how that happens.

"What is it about either this gene, or its promoter?" he asks. "We're interested in seeing whether we can track back up the signaling pathway. What's in the gene, and what turned on that signal, and that signal, and so on... We really don't know a lot about how plants sense temperature. We want to understand how we can prepare our crops."

Approach: multi-disciplinary

USDA's Plant Stress and Germplasm Development research unit in Lubbock, Texas conducts research on the response of plants to thermal and water stress and develops germplasm to reduce the impact of these stresses on crop yields and product quality. Currently, lab director John Burke is using a multidisciplinary approach in a 5-year research project to enhance plant resistance to water-deficit and thermal stresses in economically important crops. Included in the list are cotton, sorghum, corn and peanuts.

Approach: Water Use Efficiency

Plant water use efficiency (WUE) and temperature are intricately linked in many ways. Hotter temperatures, in heat waves or future climates, create the potential for more water loss from a plant. At the same time, heat waves during already hot periods lead to lower photosynthesis and productivity. Conversely, the agriculture of cooler climates may benefit from hotter temperatures, provided they have enough water. Matthew Gilbert, principal investigator at UC Davis' Whole Plant Physiology Lab, explores how water conservative plants handle heat. A plant that loses less water has a higher rate of WUE. The plant also has less evaporative cooling occurring in its leaves, and thus has hotter leaves. Apropos of climate disruption, Gilbert asks whether, when subjected to temperatures greater than 100F, a water conservative plant could be damaged by the lack of evaporative cooling.

Since all crop plants have similar physiological mechanisms of water use and photosynthesis, strategies for breeding crops with different WUE's are often broadly applicable to most field grain crops and many vegetables. Gilbert uses similar techniques and ideas to address how to change WUE in a wide variety of crops, including: wheat, common beans, lima beans, and to a lesser extent soybean, maize and almonds. "I've aligned myself with crop breeders and hope to influence their programs in the coming years. Solving these problems has to be a very collaborative endeavor," he says.

As part of the UC Davis community, Gilbert can access diverse seeds, screen them for water use in field experiments, perform scale field trials, make suggestions to active crop breeders, and collaborate on determining the genetic basis for the crop traits he finds. His broadest research objective is to decrease crop water use, while simultaneously maintaining or improving yields.

As a plant loses water through its stomata, it takes up a proportionate quantity of carbon dioxide. Gilbert hopes to capitalize on variation in this tradeoff to breed plants that use about 5 to 15 percent less water, yet still have high productivity.

Experiments in CO_2 enrichment show a plant's optimum temperature for photosynthesis increases with elevated levels of carbon dioxide. Even corn, sorghum, and sugarcane, whose photosynthesis is not stimulated by elevated CO_2, still derive some benefit of elevated CO_2 under limited water supply. The elevated CO_2 causes the stomata to partially close, which creates higher WUE, enabling these specific plants to grow for a longer time during a drought cycle. In addition, in low-temperature situations that would normally prevent or limit plant growth, enriched CO_2 stimulates growth and productivity. Of course, warming a plant beyond its maximum temperature for production is detrimental, no matter what. Bruce Kimball, a soil scientist who retired from USDA-ARS' Arid-Land Agricultural Research Center in Maricopa, Arizona, conducted FACE/T-FACE experiments on soybean at Urbana, IL, in which he and his colleagues found warming negated the beneficial effects of elevated CO_2.

Approach: OTC, FACE and T-FACE

Generally, plants grow differently in greenhouses than they do outside. With heating, cooling and CO_2 enrichment, greenhouse production can generally be much higher than outside in open-fields. Early in his career, Kimball wondered whether the increase in production at higher CO_2 observed in greenhouses and growth chambers would be true for open-field production as well. Removing the roof from the greenhouse, he created an open-top chamber (OTC) through which to blow CO_2-enriched air. The walls of the chamber confine the elevated levels of CO_2 around the plant, yet the environment inside approaches that of an open-field. However, walls still shade the plants, so the wind, humidity, and air temperature are not that of an open-field. Kimball and his colleagues working with OTC recognized these artifices and wanted to get rid of the walls and have free-air CO_2 enrichment (FACE). Engineers from Brookhaven National Laboratory created and produced the first FACE apparatus. "We did the first FACE experiment with biologically publishable results in Arizona in 1989," Kimball said in an interview in 2014.

Stephen Long began working on the effects of rising CO_2 and ozone in 1989, and began using the Arizona FACE facilities in 1992. He led the development of the SoyFACE facility at University of Illinois in 2000.

Meanwhile, Kimball conducted additional FACE experiments: two on cotton, four on wheat, and two on sorghum. The last one ended in 1999, when funding dried up. "Having more time to think and tinker, I started devising an apparatus that could provide open-field warming, so we could study the interactive effects of elevated CO_2 and warming on crops," explained the researcher. "I succeeded using infrared heaters."

Now, in addition to FACE, scientists can do T-FACE (temperature free-air controlled enhancement). About 20 experiments in different ecosystems have been done or are underway, using Kimball's T-FACE system. Research is ongoing to develop varieties that are both more responsive to elevated levels of CO_2, as well as more drought-and-heat-tolerant. Plant growth models help scientists develop strategies to maximize the benefits and minimize the detriments of the changing climate.

Adaptation & Mitigation Strategies & Modeling

Can researchers help growers produce a high quality crop and help them understand some of the dynamics that really make them a success in doing this? Hatfield said this part of the food security puzzle gets trickier as farmers have to grow new crops or new varieties of old crops for their changing climates.

Climate disruption seems to be expanding regions for various crops. The corn-growing region is expanding into the Dakotas, because it's getting warmer and a bit wetter. Unfortunately, when growing regions expand, so do areas where insects and diseases proliferate, thus impacting the growth of all crops. The direct impact of climate disruption on crop growth is relatively easy to spot. Fewer see or acknowledge the indirect impact of climate disruption on agriculture.

"I'm not so worried about climate change and agriculture," Ken Cassman said in 2014, "simply because so many of the studies that look at the impact of climate on crop yields are based on models that do not account for an intelligent farmer modifying planting dates, etc."

Cassman, the Robert B. Daugherty Professor of Agronomy at the University of Nebraska who has conducted research in Asia, Egypt and South America, expected intelligent farmers to adjust their practices in the face of climate change. Yet he questioned whether we will have enough land. Will we continue building urban expansion onto the best farmland? Will we have enough water, or pollute the water so it's not usable for growing food? Cassman and Hatfield shared the concern that simulation models cover areas that are too vast and need to be scaled down.

"One of the things that hinders us is we don't have the downscaled climate information. We really need to bring it down to the county level to be useful to producers," Hatfield explained. "That is going to require a lot of very detailed work."

To be more realistic, models need to incorporate accurate predictions for rainfall, a challenge in our changing climate. Producers need accurate information to make sound decisions, particularly regarding perennial crops. Fruit and nut growers invest 3-5 years in their trees before they produce a crop. Typically, an orchard requires a 30-year investment. To develop a refined understanding of temperature and phenology, Hatfield and his colleagues set out to improve the simulation models in the Agriculture Model Intercomparison and Improvement Program (AgMIP) by:

Completing a corn comparison with 27 different corn models from around the world,

Completing a wheat comparison with 24 different wheat models,

Gathering data from sites in a variety of ecosystems and geographic regions

Using the same data sets for each model, and

Examining how well those models predicted what was observed in each of the test sites.

At each site, scientists record the quantity and size of the leaves on each plant, the quantity of biomass, and what happens between the partitioning into reproductive structures and vegetative structures. The computer simulation model allows researchers to draw a graph of that plant as it develops over time. In comparing the models, Hatfield asked:

How much refinement does each model require to be useful in future climate scenarios?

If we're always a little warmer than the optimum temperature, what does that mean for the plant?

Are we going to over exaggerate the impact, or underestimate the impact?

If these researchers seemed to be focusing on a few major crops, that's because funding sources focus on those crops. "The pie for agricultural research needs to increase. We're not doing too much on major cereal crops, but there's not enough to go around for other crops," said Cassman. Gilbert agreed "US government investment in agricultural research has suffered greatly in the recent past. Funding for agricultural research has switched to private companies. Despite admirable success, there is a focus on particular commodities. We need much greater research investment from all sectors in solving these problems."

"Assuming increased funding, to prioritize the funding, everything must pass through a lens that says: How is this going to double (+/-) yields on existing farmland, while simultaneously massively decreasing the environmental footprint of agriculture? We're in a race against time," Cassman warned in 2014. Four years later, this is still true.

An Opportunity

It's an exciting time to be in agriculture, according to some. Others may be reminded of that old Chinese curse: May you live in interesting times. However one views the projections for humanity's food security, the information is definitely sobering.

"It's sobering, but we need to face reality," Hatfield emphasized. "There's some tremendous opportunities for us. The rewards will be tremendous, yet the collaborative aspect really needs to be enhanced. We've got to build this thing from a collaborative view point from the very beginning, instead of in our little silos."

Whether it requires snipping in other genes to change a plant's temperature response, or exploring paths in traditional breeding that change another aspect of a plant characteristic thus allowing it to better cope with the extreme weather events, researchers at the leading edge of creating food security in our changing climate have been and continue to be open to all avenues.

"A lot of us are trying to figure out how this all fits together and what it means for the stability of the food supply on a worldwide basis," Hatfield reported.

It's a true multi-sector, multi-national collaborative effort that includes agricultural professionals across the globe, including the U.S., Europe, South America, Australia, China, Japan, many African countries, India, and Pakistan. Also included are Universities, Cooperative Extension, Regional Climate Centers, and commodity groups and producer organizations. Currently the commodity groups include: soy, corn, beans, rice, and cotton. Hatfield hoped vegetable organizations would join soon. "We need to build bridges to foster communication and be able to explain what's happening realistically in terms of the crop responses."

While he was working with a geneticist to examine data for indications of different temperature responses, Hatfield knew the agricultural community couldn't expect geneticists to fix the problem. "We need a partnership," he said.

Consider how much food must be produced over the next four decades. What unforeseen environmental challenges will agriculture face in that time?

Food security is tied to political stability, to population growth rates, and to economic growth rates. The task is to find a way to produce enough food to meet the requirements of a population that's will be much larger, wealthier, and possibly stable for the first time in recorded history, if female fertility falls to replacement levels. "For 9.5 billion people, we have a chance," said Cassman. "Those 9.5 billion people have a right to flourish the way humans should. But at 10 billion people, with greenhouse gasses, etc… Really… how?"

Growers have a unique opportunity to work toward a solution to what Cassman described as the single greatest challenge facing humanity in the next four decades. Will you be one of them?

PART II

COMMUNICATING CLIMATE DISRUPTION

There's an expression, "A rising tide lifts all boats," that seems especially appropriate to the challenges we all face in living with climate disruption. Just as sailors on a ship must communicate with each other to keep it afloat, society benefits when agricultural professionals share information about climate disruption.

Blending local knowledge with scientific knowledge enables farmers and researchers to develop place-based innovations that enhance agroecosystem resilience. Opportunities for learning extend beyond the field, into conferences, workshops and even neighborly conversations. And when discussing a new topic, it's imperative to get clear on the definitions of different terms or repurposed phrases. With the shared language established, the next question becomes: how far is far enough in communicating about the connections between climate disruption and agriculture? Is it enough to talk amongst yourselves? Or do produce growers have a responsibility to share hard-won wisdom with customers or anyone who will listen?

3.
SPEAKING IN TONGUES
The Language of Climate Disruption & Ag

A shared language is an important part of getting along, and what's happening around you. In this era, when comprehending what's happening in your fields may help your neighbors to understand what's happening in their fields, and vice-a-versa; shared language around agriculture as it relates to climate disruption is, in some people's estimation, essential. This new language fosters what Laura Lengnick calls "Reflective and Shared Learning."

Lengnick is a researcher, policymaker, activist, educator and farmer. She contributed to the 3rd National Climate Assessment and wrote the book, *Resilient Agriculture* to give farmers the tools to discuss climate disruption and address its effects on their farms. "The conversation is important," she said. "As a society, we need to recognize the challenges that producers are under as a result of climate disruption. There's a legacy of denial — our leaders and consumers are able to behave as if there's nothing wrong right now and there are no new challenges in agriculture. Yet, there are."

Resilient Agriculture includes a framework used to assess the climate vulnerabilities of your operation, the exposures you're likely to experience at your location, and the sensitivities in your cropping system. As a grower, you cannot do much to change your exposures, but you can do a lot to change your adaptive capacity and sensitivity. The terminology presented offers farmers a new way to think through and manage challenges created by climate disruption. A shared language connects you to researchers and other agriculture pros, and helps everyone collaborate and figure out what to do about climate disruption.

Ann Adams, Executive Director of Holistic Management International, expressed hope that a shared language may lead to improved policies and incentives that will inspire you and your colleagues to adopt practices that lead to the mitigation or reversal of climate disruption. In Adams' estimation, creating greater resiliency in U.S. landscapes will enable U.S. farmers to feed U.S. consumers. She said that will, "decrease dependency on foreign food over which we have little to no control."

S. Elwynn Taylor, the extension climatologist and professor of Agricultural Meteorology at Iowa State University, said it's key for the entire agricultural community to understand four concepts.

Weather Risk Management means knowing your likely exposures- the risk of flood or drought or out of season heat wave or chill - and the impact of those exposures on crop and market price (Lengnick calls these your farm's Sensitivities).

Ag practices have contributed to the changing of the composition of our atmosphere.

Agriculture need not be a part of the problem. The better way is also more profitable in the short run to the farmer.

There is no known solution to the increasing CO_2 in the atmosphere that does not include agriculture.

The emergence of a shared language around agriculture and climate disruption also facilitates the development of a common vision for a truly sustainable agricultural system.

In Jimmy Bramblett's view, a shared language equals a shared vision. As Wisconsin's Chief Conservationist with the Natural Resources Conservation Service (USDA-NRCS), Bramblett focused on the relationship between soil health and climate disruption (Today, he is USDA-NRCS' Deputy Chief of Programs in Washington, D.C.). In an interview in 2015, Bramblett said that a common frame of reference enables more stable production with less risk from the direct and indirect impacts of climate disruption.

As the nation's primary authority for soil mapping and soil data collection, the NRCS has been leading a highly successful soil health campaign for the past several years. Many agricultural professionals are also communicating in this shared language.

-"Resilience through nimbleness in management decisions and a generational transformation works best," said Thomas Harter, Cooperative Extension specialist in the Department of Land, Air and Water Resources at UC Davis.

As the language of climate disruption and agriculture continues to develop and emerge, you're likely to hear it more at conferences and events organized by extension specialists and other organizations that support farmers; and to see it in more media. Harter, who organized a conference focused on protecting groundwater resources in California in 2016, said many "aha" moments resulted from attendees listening to one-another and learning each other's language. "The shared experience at the conference demonstrated that there is much momentum and interest in agriculture to protect groundwater resources and quality," Harter said.

Often the same words have different meanings to different people. Paul West, co-director and lead scientist of the Global Landscapes Initiative at the University of Minnesota's Institute on the Environment, explained why agriculture needs a shared language to talk about climate disruption. "For me, resilience means the ability to adapt and still thrive in new situations. But that's obviously hard to quantify... The specifics will need to change depending on what the areas need to adapt to."

Molly Brown, Associate Research Professor at Department of Geographical Sciences at University of Maryland, explores the use of long-term records of vegetation, rainfall, soil moisture and evaporative stress in agriculture insurance programs in Africa. She said developing a universal language around farming and climate change is a critical first step in recognizing the real objective of agricultural research and development.

"If you're a first-world scientist and you go to a conference and learn about the concerns of the farmers in the developing world, you may not change your research, but you can change how you express the meaning of that research in your own papers to these other communities," said Brown. "It's only through coming up with a universal language. You need to be aware that everywhere does not look like Kansas, and that the objective of ag development is not to make everywhere like Kansas. It's to have everyone achieve their own capabilities and their own needs and desires. It's not to make a cookie cutter world."

The Emerging Shared Language of Agriculture and Climate Disruption: *a glossary*

These following terms have evolved to help farmers and researchers communicate about climate disruption.

- *Climate-smart Agricultural System*: a farm that is designed to operate and even thrive through the extreme conditions caused by a changing climate.
- *Climate Resilience / Adaptive Capacity:* both terms refer to a farm's ability to thrive despite the challenges created by climate disruption.
- *Soil Carbon Sequestration*: the process whereby root systems store much of the carbon dioxide that crops pull from the atmosphere deep in the soil.

As is clear from this list, many of the terms emerging as part of the shared language of farming and climate disruption are words and phrases pulled from everyday language and adapted to fit the specific needs of the field (no pun intended). Other terms from resilience science include:

- *Response capacity*: a farm's ability to respond quickly and effectively to avoid or reduce damages from disturbances;
- *Recovery capacity*: a farm's ability to quickly recover post-damage;
- *Transformation capacity*: the farm's ability to transition to a new type of system when necessary.

Jerry Hatfield, co-lead author for the agriculture section of the 3rd National Climate Assessment developed an equation that stresses the importance of managing (M) one's farm system to optimize the performance of the genetics (G), given all of the variations in the environment (E):

$$G \times E \times M$$

Hatfield's concept of Genetics x Environment x Management helps many growers see the larger picture and how the pieces fit together.

4.
TO TEACH OR NOT TO TEACH
Educating Customers

In the U.S., the average consumer is trained to shop for the best deal.

Since Congress passed the Organic Foods Production Act in 1990, organic farmers have needed to educate consumers about the personal health and environmental benefits of organic growing methods in order to help people justify the higher cost of organic produce.

Now, as climate-smart farming practices are spreading, it may be necessary to educate consumers again to help them justify their purchases. I asked farmers across the U.S. to share if, how and why they teach their clientele about climate disruption.

Their responses varied widely.

Julie Rawson, of Many Hands Organic Farm in Barre, Massachusetts does outreach to CSA members and the farm's 1,000-person information list about carbon sequestration and climate disruption.

"I live with a low-grade anxiety about climate change and see it pretty regularly manifested in the weather with its impacts on the farm," she said. "As a farmer, I believe I have an important but miniscule part in changing the climate trajectory with our farming practices."

Rawson educates her clientele by regularly talking about practices that support good carbon management on the farm or garden. She encourages people to try these climate smart growing practices on their own piece of land.

"When our food quality took an upward leap last year I 'blamed it' on carbon sequestering techniques that we are learning to use on our farm," Rawson explained. "The customers noted an improvement in quality and taste, so it was easy to make that connection."

In Wauzeka, Wisconsin, Renee Randall of Willow Ridge Organic Farm, saw four decades of careful land stewardship decimated by a tornado in 2015. She has also experienced growing seasons devastated by hail, hard and pounding rains, cold and "record-breaking" wet spring flooding. Conceivably, her customers would also have experienced some of these weather extremes and made the connection to climate change, but Randall said trying to extend that reality to the customer is much more difficult.

The CSA concept is to connect consumers with farmers by giving them access to the farm and an experience of nature through being a shareholder. Hoophouse growing, hydroponics and aquaponics are methods that shield crops from climate disruption in order to make food consistently available in all conditions.

While it is a definite bonus to have access to farm fresh food in any weather, Randall said the constant availability of fresh produce takes attention away from the effects of extreme weather events on field growing, where weather is a definite factor. Although her personal experience would seem to be the perfect springboard for climate education, Randall found it difficult to attribute it to climate disruption, since there was so much debate about why the weather was off. "People thought of weather problems as excuses that should have a quick fix. I had more than one weather disaster and with each tried to show the effect extreme weather and changing weather patterns have on farming. No matter how beautifully I wrote about the fallout after the tornado, my customers mostly wanted to know when they could expect a delivery."

Marko Colby and Hanako Myers, of Midori Farm in Quilcene, Washington do not intentionally educate customers about climate disruption. When the subject of crazy weather patterns does come up, he usually just says it's climate *chaos*. Colby doesn't feel like it's his role to educate his customers about anything, and doesn't want to burn bridges with customers who consider climate disruption a political, rather than a scientific, issue.

"They buy from us because we grow high-quality produce, not to hear my opinions about what I presume is going on in the world. We really make it a point to be non-political, non-confrontational, in all of our outward farm marketing / interactions," he shared. "We want to bring folks into the fold of enjoying fresh local produce regardless of their world view and or political beliefs. Once they realize how good local organic vegetables are, they may dig deeper and try to understand humans' role in producing greenhouse gases."

At Stamford Museum and Nature Center (SMNC) — a working farm in Stamford, Connecticut — Education Director Lisa Monachelli said climate disruption education vis-a-vis agriculture is an important part of their work. SMNC offers a program for schools and discussion about climate disruption also arises regularly in the nature center's other programs, most notably Maple Syruping.

"We talk with the visitors about how Southern Connecticut's maple trees and resulting syrup production are affected by climate change," Monachelli said. "To make the best decisions for our environment, all of us need connection with the natural world, as well as an understanding of the science behind its processes and how human impact can affect those processes. Naturally, product yield is important in farming and the causes of yield variability are part of the educational process."

Ridge Shinn and Lynne Pledger, co-owners of Big Picture Beef, in Hardwick, Massachusetts have built an entire business model around developing livestock production methods that fix the water cycle and sequester carbon. "Our very name was chosen because we want people to understand it isn't just about taste, it's also about the larger issues," Pledger said.

Shinn has been raising grass-fed beef for fifteen years and has a forty-plus year history with both dairy and beef cattle. He and Pledger founded Big Picture Beef in 2016 and developed point-of-sale materials to educate consumers the grocers who carry their product (Currently, the Big Y chain of grocers is their biggest client.).

They also educate consumers with public speaking engagements and a highly informative website that includes information about the connection between rotational grazing on and bigger yields of crops.

"We feel the climate/food connector is critical to explain to consumers so they can make informed choices," Shinn said. "The correct choices help mitigate climate/carbon issues and reinvigorate the local economy."

Whether offering education about climate disruption and climate smart farming practices will benefit your business or hinder it is something only you can figure out, based on your market. No matter what you decide, "The deliciousness of good food is one thing that brings people from all walks of life together," Colby said. "That is more important now than ever."

PART III

DISASTER AT YOUR DOOR

METHODS FOR CLIMATE RESILIENCE IN EXTREME TIMES

Farmers have always coped with calamities like drought and flooding. But in recent years, superstorms have increased in frequency and intensity. Such extreme weather events seem like the stuff of movies until they happen to you.

Are you prepared for misfortune? In this section, farmers share their stories of disaster and recovery, and experts offer resources to help you plan for the worst.

5.

LESSONS FROM THE FIELD
How Proactive Farmers Are Resilient in Extreme Weather

 severe hail storm swept across Bob Quinn's Montana ranch in early August 2015, long after the hail season was over. Quinn was concerned, but did not panic. Hail storms typically last a few minutes, but this storm lasted over 20 minutes and did enormous damage to crops, even though the hailstones were small. However, Quinn knew his planting strategy was such that the storm wouldn't cause a total loss.

"Our crops are so diversified, the damage to each different crop was quite variable," Quinn said. The soil-building green manure legume crops had already terminated and been worked into the soil, so there was no loss there. The alfalfa hay had already been swathed and baled, again escaping loss. The safflower was blooming, but Quinn said the plant is so sturdy that there was no visible damage immediately after the storm. (A few darkened kernels were seen at harvest time a couple months later.)

The grain crops did not fare as well. Quinn's Kamut brand khorasan (an ancient spring wheat) suffered an approximate thirty-five percent loss. Because the Kamut was not yet ripe, many kernels were still soft and green and not easily knocked out of the heads. In contrast, the winter wheat kernels, which were ripe and nearly ready to cut at the time of the storm, were easily knocked out of the head by the hail stones. Quinn lost about seventy percent of that crop. The feed barley fared worst. "Its heads were erect and the kernels fully exposed to the hail," Quinn said. "The hail was not big enough to break the straw which normally happens, but every single kernel was stripped off, so the crop was a total loss. The peas, which were also a quite near to ripe were easily shelled out by the hail and were also a total loss."

In the winters both before and after the hail storm, Quinn experienced several extra warm SW winds (chinooks) than usual. Each chinook was followed by sudden and extreme cold, which Quinn said is also unusual. In 2014, Montana had abnormal highs in the 60's in late November. This was followed by a sudden storm that dropped temperatures by 50 degrees in 36 hours. These extreme temperature swings in winter 2014 and 2015 killed or damaged over half of the twenty-three different varieties of semi-dwarf apple trees in Quinn's experimental orchard. "The varieties I had planted were all rated for zone 3, meaning they would survive temperatures of -40 F in the winter," he said. Quinn attributed the loss of his apple trees, which never experienced temperatures of -40F, to rapid and extreme swings in temperature and rapid fluctuations in moisture during the dormant season of late fall to early spring.

"Looking at a longer time span, we used to experience approximately eleven-year cycles of dry and wet years. Following the wet years in the mid-1990's, it was twenty years before the next next wet year. During that time, we had two very dry periods. This year we had a drought following the wettest year we have had in twenty years."

These dramatic weather patterns exemplify the types of weather fluctuations that have become more common throughout the world in the last decade. In Alaska, farmers Susan Willsrud and her partner Tom Zimmer have observed radically different frost dates, planting times, precipitation, and temperatures from year to year. Their strategy for dealing with highly unpredictable weather shifts, which they attributed to climate disruption, is the same as Bob Quinn's: Diversity.

"We grow a really large range of crops - so each year something will do really well under the conditions and other things won't, but it all works out fine," Willsrud said.

Willsrud and Zimmer grow 40-50 different crops (including herbs, vegetables, and cut flowers) each year. "We aren't going down the road of specializing in our higher value crops, knowing that we could have unexpected conditions," Willsrud said. "We are avoiding larger amounts of crop loss by continuing to grow many crops -- also we are keeping a high number of our market in local CSA."

Willsrud and Zimmer appreciate the diverse nature of CSA's, both in number of crops and in the number of individual consumers. They feel the CSA model offers farmers more relationship as well as more stability. "It works fine to have some people come and go each year. Also, in a CSA people are investing in the farm and also sharing both the risk and the bounty; with unexpected weather, this is a benefit," said Willsrud.

In any system or organization, from organic agriculture to businesses, to community action groups, to CSA's, diversity begets stability. Quinn believes this principle applies in the short term and the long term. "In addition, organic agriculture provides one of the greatest potentials for mediating climate change by creating an enormous carbon sink via carbon sequestration which happens when green manure crops are turned back into the soil," he added. "Also, the production of tremendous amounts of greenhouse gasses is avoided when the production of chemical fertilizers, not needed in organic systems, is first reduced and finally stops. These two things alone would completely reverse the growing accumulation of carbon dioxide and other greenhouse gasses."

Hand-in-hand with choosing diverse crops is carefully selecting crops (and animals, if you grow livestock) that are appropriate for your zone, your microclimate and the types of weather extremes you are likely to experience. Willsrud and Zimmer raise heritage Shetland sheep, a resilient breed that thrives under a wide range of conditions. The farmers selected them for extraordinary fiber quality, overall vigor, and mothering ability.

Quinn finds older selections of crops are more resilient, because these varieties are inherently diverse. "The older lines are not pure lines because they were originally collected in the wild; they have many strains which are closely related. This is unlike the monoculture pure strain resulting from modern plant breeding and selection. The mixture of multiple strains, called a land-race, results in a crop which has [individual plants with] different resistance to extreme weather, insects, and disease. Therefore, when subjected to any of these challenges many of the resistant plants will survive while some of the susceptible plants will not. The resistant plants can then grow bigger and fill in the spaces left by those plants which died."

In contrast, when a pure line is hit by extreme weather, if one plant is susceptible, they are all susceptible, and the whole crop will be lost, leaving no survivors to fill in the gaps. In Quinn's experiments with apple trees, he found that standard trees are more resilient than semi-dwarf trees. He has concluded that the rootstock used to create the semi-dwarf also creates stresses, which lowers the trees' resistance to weather extremes. "As the weaker semi-dwarf trees die or are severely damaged from extremes in fall, winter and spring weather, I am replacing them with standard trees," he shared.

6.
NOW WHAT?

Across the U.S., farmers have been experiencing disasters at an alarming rate. Here, five producers share how they recovered.

In early 2012, Dayna Burtness launched her first farm on 6 acres of rented land in Northfield, Minnesota. She got off to a great start, landing high-profile restaurateurs from the Twin Cities as clientele and creating success in part by seeking her clientele's suggestions about which produce to cultivate. Then, in June of her first season, disaster struck. Eight inches of rain created a river 50-feet wide that drowned a third of Burtness' crops and flooded her rented greenhouse space with 4 feet of water.

"That flood has affected most of the farming decisions I've made since," said Burtness. Ultimately, the farmer opted to scale back on annual vegetables. Today, she maintains a quarter-acre market garden for friends and family on a 67-acre farmstead in Spring Grove, MN, and focuses on raising pastured livestock. "They're way more resilient," Burtness said.

Double Whammy

In a good year, the Rogers family harvests between 2500 and 3000 pounds of almonds on their 175-acre ranch in Madera, California. Founded over 100 years ago, the first almond trees were planted in the 1980s. By 2003, the third generation of Rogers brothers had switched their entire operation to almonds. In 2011, the California drought started. Farming a thirsty crop in a drought was challenging enough. Then disaster hit two years in a row.

In 2014, during their most important growing period, the Rogers' well stopped working. They still had water in the well but couldn't access it until the well was fixed. It took five weeks for the parts they needed to arrive. In

May 2015, shortly after the fruit had set, a hailstorm destroyed approximately 15 percent of the almond crop. Tom Rogers estimated the losses in one field were closer to 40 percent. Hailstorms are not abnormal for central CA, but that storm was the first of that magnitude that the Rogers remember. "We had never seen that kind of a loss in almond," said Tom. " It looked like someone had come and shook the almonds off the orchard."

Because the hailstorm injured the trees, the Rogers had to alter their management techniques for about three weeks. It's a documented response that when a tree goes through a major stress, like a hail storm, it sheds more fruit about three weeks later to make itself healthy. To prevent this response, the Rogers changed the way they watered and fertilized - applying foliar products with microbes to essentially tell the tree, "You're okay."

Although the yield was down, the trees did not drop more fruit after the hailstorm. After the crisis passed, the Rogers assessed what they were doing and found that time spent in management went way up, while the time spent in the field decreased.

Losing their well during their important growing period led the Rogers to start using pulse irrigation on the ranch. They put soil moisture meters and weather stations all over the farm to monitor patterns, gather information, and help them make the best decisions. Now, Tom Rogers sits down at his computer each day to program the irrigation. The goal of pulse irrigation is to meet the needs of the tree without sending water beyond the root zone. The Rogers water one hour at a time, 3 times a day, seven days a week. "By pulse irrigating, we've reduced the amount of water that we're using," said Tom, who estimated they were applying 25 percent to 30 percent less water in 2016 than they were prior to losing the well. In addition, the almonds appeared to be doing better in 2016 than before. "We're excited about that; we're still learning; we'll be a lifetime figuring it out."

Despite the disasters they faced in 2014 and 2015, the Rogers maintained production close to 2500 pounds. "When you get backed into a corner, you start looking at: How can I do this better?"

Tornado

"Everything was whole and then the fabric of everything you've done is torn apart. You don't know where to start. You don't even feel like starting again. It's such a mess."

~ Renee Randall

On Monday, June 22nd, 2015, a severe storm hit the tri-state area of Iowa, Minnesota, and Wisconsin. Renee Randall, of Willow Ridge Farm in Wauzeka, Wisconsin, was on the phone when it started.

"I was looking out the window and it exploded into the house. Shards of glass were everywhere. I went to the basement. When I came up, everything was completely different in my life. I have a two-story farm house. Part of the roof was gone. It was pouring rain into the basement. Wires were hanging everywhere. A giant maple was going across my porch. I couldn't get out of the front door. This is not tornado alley where I live. I never expected a tornado."

Randall lost almost everything when the tornado whipped across her 117 acres. The hoop house was crushed. The storage shed was blown off its foundation. The pounding rain, hail, and high winds took a severe toll on the fruits and vegetables in the fields that weren't leveled by the tornado. Trees were downed.

Renee Randall experienced the kind of disaster that doesn't just affect a person - it traumatizes them. Over a year later, Randall was still feeling the effects and wasn't sure she'd be able to restore the operation and the life she had spent over forty years building.

Ninety-one landowners and over 3000 acres of privately-owned woodlands were affected in Crawford County, WI. Crawford County did not meet the economic threshold for FEMA assistance, but Randall believes that, had the damage to woodlands been included in the numbers, area landowners would have been eligible. The County Executive Director for USDA's Farm Service Agency (FSA) was looking for the kind of crop damage he was familiar with- corn and beans. He had not considered trees a crop, so never assessed the damage to the woodlands as a farm loss. "In our area, farmers have used their trees as almost a savings account," said Randall. "During the farm crisis of the 80's, logging was used to offset debt."

In 2015, Department of Natural Resources (DNR) estimated Crawford County's damaged timber might sell as salvage, worth 30 percent of its original value. When, as a special request, the DNR forester examined Randall's woods, he found that every acre of the 40-acre forest had tree damage.

"Twisted, uprooted, broken mature oak trees is not the kind of damage a farmer can handle without putting their life on the line," said Randall, who felt hopeless and frustrated until she learned about the Emergency Forestry Restoration Program (EFRP). The program had never been used or even offered by any agency in Wisconsin or surrounding states. "It's on the books--legislated--and never offered out to farmers," Randall said. "When I mentioned that it was never offered out, the response was that it was up to farmers to know what's on the books."

The passion Randall brought to farming is now directed at helping her neighbors to recover some of the loss via EFRP. She started by lobbying FSA's county committee to approve the necessity for it.

"I've worked on it since January. It's gone to Washington with the request for EFRP funding. It's been approved at that level, but no money has been allocated yet. Because it's a cost-share program, you have to spend the money first and then get reimbursed by the government," she explained. "And, since this program has never been implemented it's taking some time to administer. Now that the program has emerged as a given entity, the county director has thrown himself into making this program workable."

Randall will get money to clean up and restore her woods. However, the oak woodland she stewarded for decades will take over fifty years to come back. Randall will not live to see the result of her efforts to save her forest.

To recover her produce operation, Randall leveraged relationships. Members of her CSA supported her through the tornado, and King's Hill farm donated produce to help her through to the end of the season. Friends created a GO FUND ME page that raised about $4500. The funds helped her clean and replant and enabled her to finish the season. Neighbors bartered with

Randall: organic hay in exchange for a new metal roof on her house. Seeing how depressing the landscape damage was in front of the house, they donated a day to give the landscaping a face lift. "It was a spirit lift as well," she said.

Randall reported a strong 2016 CSA signup. At the time of this writing, she was managing to supply the CSA, despite the fact that she still hadn't been able to replace her hoop house, shed, and small equipment. "If I can make it through this year, I'll have to reassess what needs to be done next," Randall said.

Flood

In 1999, Vicki Westerhoff took over her family's 20-acre farm in St. Anne, Illinois and established a farming partnership called Genesis Growers, Inc. By 2015, Genesis Growers was a 65-acre certified organic farm with sales at farmers markets, wholesale to restaurants, and a CSA. Westerhoff was growing approximately 350 varieties of herbs and vegetables, until a six-week deluge in June and July 2015 flooded her fields and destroyed her crops.

"We dug ditches to drain the water repeatedly," said Westerhoff. "But rains kept falling. I would get the water drained off and we would have another deluge. It rained several inches at a time and in close enough proximity that we never really dried out. The final flood ran like white water rapids across my land. At that point, any crop that had managed to survive was wiped away."

Westerhoff altered her cropping scheme and tried field leveling to facilitate drainage after the flood in 2015. "Next year I am going to do more work in this area and will implement a raised bed system. Since I have very light soil, I have not felt the need to use raised beds, but when the rains come unbidden they would help," she said.

The deluges of 2015 ended in late July. Westerhoff started re-planting fast growing baby greens and root vegetables on July 28. With a lot of hard work and diligence (and the delayed onset of cold weather) Westerhoff and her team managed to harvest a decent crop, which she sold over the winter months. "Farming is a tough business with low profit margins, so a disaster can cause incredible problems," said Westerhoff. "So far this year we have made it, but [we] are by no means in good shape financially. It would be a struggle no matter what, but we began having continual rains in July this year that have damaged many crops. I am back to having to replant in hopes of a good winter season to tide us over."

Westerhoff described herself as a fortunate farmer who received help from loyal customers. "From the beginning I decided I did not want to be a nameless, faceless farmer. The fact that I know so many of my customers is the reason I am still able to farm and I thank them all with all my being."

When Disaster Hits

Responding to a life-altering disaster requires the ability to think quickly and spring to action, the willingness to reach out for help and find resources you may not have known existed. When an event occurs that brings you to your knees, perhaps the most important tool for survival is adaptability - a willingness to learn from the experience, to learn from others, and to envision a new modus operandi for yourself and your operation.

Ana Otto, of Arizona Farm Bureau, advises farmers to have cash reserves and a good relationship with their local loan officer. Farmers lacking cash reserves and sufficient credit can turn to various aid programs. In 2014 and 2015, the CCOF Foundation provided $33,000 in assistance to organic operations from 18 states, including Willow Ridge Organic Farm and Genesis Growers, Inc. In addition to CCOF's $500 Bricmont Hardship Assistance Award, Randall received a Farm Aid grant for $500, to be used for household (not business) expenses.

Farming organizations are another valuable resource to turn to in times of distress. "When natural disasters occur, Farm Bureau can help lobby state and federal officials for disaster relief if it is slow in coming," Otto said.

In addition, a national community of farmers and ranchers within Farm Bureau provides financial and other assistance when disaster strikes. In August of 2016, after floods in Louisiana caused an estimated $100M+ in crop losses during harvest time, American Farm Bureau reached out to Louisiana Farm Bureau to offer assistance.

Relationships with suppliers, crop specialists, extension agents, etc. are another key to getting back on track after disaster strikes.

Burtness, Randall, Rogers and Westerhoff recovered from the disasters that befell their farms, but their operations will be forever changed. In some cases, for the better; in other cases, it's too soon to render judgement. Burtness took the simplest approach: changing focus. Randall received assistance from friends, neighbors and private and public funding programs. Until her farm is cleaned up and safe again, she can only grow a fraction of the produce she once did. The Rogers found learning to be their salvation. Westerhoff reached out for information and accepted help from the supportive community she had spent seventeen years cultivating. What can you do if faced with a disaster?

Advice For Tough Times

"Be adaptable. Look at what's out there. After the last two years, we look at all new technology. Talk to whoever's doing something interesting. Try something different. The reality is we're gonna be forced to. We have to adapt. For us, this is an exciting time in agriculture."

~ Tom Rogers, almond grower, Madera, CA

"Be prepared as best as one can. Be flexible and able to put a new plan into action if need be. Last year I made it all the way to Plan E before I was able to plant. Plan E worked. I could have given up. But, it was essential if I wanted to survive to continue trying."

~ Vicki Westerhoff, Genesis Growers, Inc., St. Anne, IL

"Look to yourself as the best resource; you're a farmer. Don't give up, don't be dismissed. It's only because I pushed and figured things out that we've gone this far with a government program that never would have seen the light of day otherwise."

~ Renee Randall, Willow Ridge Organic Farm, Wauzeka, WI

"Never give up. Request FEMA and other natural disaster assistance that may be available, contact your local NRCS office to assist financially and technically with conservation practice planning and implementation for recovery. Contact the USDA Emergency Watershed Program (EWP) and the Emergency Conservation Program (ECP). Call a special meeting with all the agriculture interest groups and local, state, and federal agencies to collaborate a strategy moving forward to provide necessary assistance utilizing each other's resources to the maximum."

~ Curtis Elke, State Conservationist, USDA Natural Resources Conservation Service (NRCS), Idaho

"Contact your local NRCS office first. Avail yourself of all NRCS services including natural resource assessments, planning and design services as well as financial programs to help repair and/or enhance/ conserve the resources of your farm that will enable you to keep your farm productive and thriving into the future. NRCS is a great one stop and first stop service. NRCS administers various Farm Bill programs that provide money for remedial practices to address damages caused by disaster or natural event. NRCS provides free non-regulatory technical planning and design assistance for farms impacted by natural events, including above normal rainfall events that might only cause minor damage to fields or crops. NRCS works closely with Resource Conservation Districts, which in turn provide additional free assistance, like help with permits in some counties, other technical and/or financial assistance. NRCS also works closely with the USDA Farm Services Agency to help growers with emergency and disaster programs, funding, and help with crop losses."

~Rich Casale NRCS District Conservationist, Santa Cruz County, CA.

"Many Holistic Management producers have found that investing in soil and land health is the number one best way to build land resilience for future disasters and more quickly recover from these kinds of disasters. Improved ground cover has helped people be less susceptible to flooding, improved grazing practices have helped people cover ground after a fire or reduce fuel loads to reduce likelihood of fire in the future, and soil fertility practices have helped farmers and ranchers alike to mitigate the effects of drought and pest invasion. Lastly, the Holistic Management decision-making and planning processes have helped producers prioritize decisions that must be made after a catastrophic event. Free e-books on their decision-making, financial planning, and grazing planning are at: https://holisticmanagement.org/free-downloads/. As one rancher noted, 'I had a wildfire burn 1/3 of my ranch. Within 24 hours I knew what my plan was and how I would deal with. That's peace of mind.'"

~Ann Adams, Executive Director of Holistic Management International in Albuquerque, New Mexico.

7.

PREPARE FOR THE WORST
Tips for Recovering Emotionally, Financially & Physically

I n 2012, David Crafton moved to Norway, South Carolina, to farm full time. Norway is a town with fewer than 200 households in Orangeburg county, where row crops dominate agriculture; and vegetables, fruits, and nuts are regarded as "Specialty Crops" by the state Department Of Agriculture. Crafton raises pastured pigs and grows a market garden using methods he calls, "beyond organic."

In 2014, a drought hit South Carolina that lasted two years and was followed by a flood in the autumn of 2015. Local media declared 2015 an historic crop year for Orangeburg county, because of widespread losses. Crafton lost his entire vegetable crop, as well as the pasture he had planted to feed his pigs throughout the winter. The toll of the losses extended beyond financial. Crafton experienced emotional distress, as well.

People, especially pick-yourself-up-by-the-bootstrap types, tend to avoid discussing emotional effects of a disaster. Yet, stress, grief, anxiety, and depression are common side effects of major disasters. According to the American Psychological Association (APA):

Even when you're not hurt physically, disasters can take an emotional toll. Normal reactions may include intense, unpredictable feelings; trouble concentrating or making decisions; disrupted eating and sleeping patterns; emotional upsets on anniversaries or other reminders; strained personal relationships; and physical symptoms such as headaches, nausea or chest pain. Psychological research shows that many people [can] successfully recover from disaster. Taking active steps to cope is important.

Vivian Marinelli is a psychologist and senior director of crisis management services at FEI Behavioral Health in Milwaukee. FEI provides resiliency solutions and crisis management services to communities and organizations throughout the nation. Marinelli said preparing physically and emotionally for disaster can help people cope when disaster strikes. There are four phases in what FEI calls the "disaster management cycle":

- Mitigation,
- Preparedness,
- Response, and
- Recovery.

Mitigation involves doing a risk assessment and considering the types of disasters that typically happen in your area, as well as personal events that would wreak havoc on your operation. On a large scale, these may be extreme weather events, fires, or explosions. On the personal side, it may be a sudden illness or death of a family member or key farm staff. It may be a blight or disease that causes total crop loss. Marinelli advised conducting an overall assessment for your farm. It can also be helpful to learn about risk assessments done by and for the community.

Preparedness essentially means identifying resources that one can access in a crisis. Do you have all the resources you need on the farm, or do you share certain resources with neighboring farms? Does your community have sharing agreements with other communities? In Waukesha County, Wisconsin, county officials embrace an all-hazards planning approach, since it is impossible to predict all aspects of a disaster situation. "This planning benefits all Waukesha County residents, both those of the farming community and the community at large," said Bridget Gnadt, of Waukesha County Government.

In your preparedness plan, identify who has what resources, who to call, and emergency contact numbers. Part of preparing is developing relationships. It's easy to get caught in the sunup-to-sundown farming schedule, but Marinelli stressed the importance of taking time to reach out and connect with people who might be able to offer support in a crisis.

Marinelli also advised considering all aspects of the worst-case scenario and planning accordingly. For smaller weather-related crises, make sure you:

Have some way to access weather reports in the event of a power outage (e.g. short-band radios);

Fuel all generators;

Charge all cell phones, laptops, and other battery-operated communication devices;

Stock a first-aid kit with basics as well as prescription medicines your family members require. For example, if you have diabetes, make sure you have insulin and can keep it cool.

Marinelli also suggested moving equipment away from the buildings and tying it down. If high winds or a tornado comes through, and equipment flips over, the likelihood is you'll just have to turn it, rather than fix a damaged building because of it.

Finally, get to know your community resources, including the locations of local shelters. Waukesha County plans ahead to address immediate concerns, like temporary feeding and sheltering of people and animals. As part of their "all-hazards approach," the county recognizes the importance of planning to address the needs of all people, including those with access and functional needs. They also conduct planning to address the psychological needs of people in disaster.

What you can do to prepare emotionally for disaster is consider the emotional support system you already have in place. Is it family, people at your favorite coffee shop, or your pastor/rabbi/imam? Consider what you do on a day-to-day basis to cope with stress. Is it taking a walk on the property, listening to music, reading a book? Getting into the daily habit of relieving stress will give you the emotional tools you need to stay calmer in a crisis.

Response depends on the level of disaster. State or national agencies like FEMA may offer support. "One of the biggest issues we've seen following a disaster is communicating," said Marinelli. "How do you reach people when phone lines are down, roads are blocked?"

Marinelli said she has observed farmers' reluctance to use federal resources during major disasters. FEI encourages agencies and organizations to send communications in the various languages that are spoken in the community. FEI encourages Farmers who are uncomfortable seeking help from state or federal agencies for any reason to turn to local community agencies, including social organizations and churches.

"It really is about personal planning," said Marinelli. "Develop your own emergency response plan as a farmer. You don't just have your individual needs, your family needs; you need to support your farm, as well, especially if you've got livestock, because they're depending on you taking care of them." (See the Farm Emergency Preparedness Plan checklist at the end of this chapter).

If flooding or downed trees block roads, call the people on your list, whether they're a farmer down the street, a farmer in another county, or a family member in a different area code from you. If the area code where the storm occurred is not available, you need to be able to reach someone and say "I'm trapped. I'm okay, but roads are blocked and I need some help." Because they're in a different area code, they can let relief workers know who you are, where you are, and that you need assistance.

Responding to Emotional Upheaval

If you reach the point where you need a more than your daily stress-relieving routines to feel whole, local community social service agencies may offer counseling for individuals or families. Some agencies offer support services for a very low cost.

In a large-scale disaster, the Red Cross usually offers support, even referring individuals and families to counseling. The American Psychological Association reports that, while psychologists are often mobilized to help at disaster sites, they do not offer therapy at disaster sites. Instead, psychologists help people experiencing disaster, "to move from feeling hopeless to having a more long-term, realistic perspective. This process can include taking small steps toward concrete goals and connecting with others as they learn to cope with a disaster's logistical and emotional challenges."

"It really is taking a look at what do I do day-to-day," said Marinelli. "Where are my needs? When things get really rough, what support systems do I have? Probably the first and most difficult piece is to even identify that [you] need additional help."

For Crafton, the emotional toll of the 2015 crop year was steep. "The emotional toll had everything to do with the financial toll," he said. During the drought, he watered his crops but 100-degree temperatures made it hard to keep moisture in the soil. "There was no insurance to cover what I was doing. My typical monthly income shrunk by $2000-$3000. It doesn't sound like a lot, but it's a lot ot me. [After the deluge,] my fields were flooded for weeks."

According to Crafton, South Carolina's response to the drought/deluge cycle in 2015 was lacking. "We have a lot of disorganized organizations," he said. In contrast, Waukesha County carefully conducts damage assessments and meticulously documents response and recovery costs to increase the chances of qualifying for state or federal disaster aid. The types of services offered vary depending on the type of disaster, the stage of the disaster response, the magnitude of the event and

> ## HOW TO CONDUCT PSYCHOLOGICAL FIRST AID IN A DISASTER
>
> Effective emergency preparedness and response is a community effort. Someone who has just experienced a disaster does not need professional psychological counseling, but does need help and understanding. When someone has just experienced a disaster, you can help minimize the negative emotional impact of the situation by following these steps:
>
> 1. Establish a caring, compassionate connection with the affected individual,
> 2. Provide physical & emotional comfort,
> 3. Provide as much information as possible about the response so that everyone understands how it is being coordinated as well as the next steps for them in the recovery, and
> 4. Empower survivors to address their own needs by connecting them to agencies and resources that may assist with both immediate and ongoing needs.

the availability of aid. Once all immediate subsistence needs are met, Waukesha officials link affected individuals and families to additional community resources to help in long term recovery of all types.

Crafton considers himself lucky that his house and barn didn't flood, because of their hilltop location. "Some people lost their homes. The pigs survived in the barn." However, since the flood destroyed the forage, Crafton had to feed the pigs hay. Purchasing food for over 100 pigs added financial strain.

To get through, Crafton cut corners wherever he could to get through the year and hoped for the best. "When you have something like that, and it pretty much destroys your forage and your crops, it takes years to recover when you don't have insurance," he said. "The money that I didn't make the previous year would have typically gone into seed for the following year, fuel, equipment and whatever else I needed, including food for the pigs. Growing seasonal crops, you wonder how things are gonna be for the next year. You cut corners in your personal life, to keep the farm going. You eat cheaper. You don't travel."

Before the disaster, Crafton was living with his girlfriend and her two children. "I had a family," he said. "The stress of the events had a lot to do with the end of the relationship. She worked from home. The kids helped out on the farm. Financial stress is hard on anybody, but when it takes years to recover, it's just that much harder for everyone. It's like, you get all the typical things you would have in any relationship and you gotta multiply it, because that toll isn't just for a month or two; it's a long, enduring emotional roller coaster, and it's a hard roller coaster to ride."

Recovering Financially and Emotionally

Crafton filed for help from an independent regional bank with deep roots in the community. They sent him a check for $500. Crafton also reached out to South Carolina Farm Aid. They sent a check for $250. While the $750 in aid did help, it didn't come close to covering the cost of Crafton's losses.

"The economic impact cannot be ignored," said Gnadt. When farm damage occurs, Waukesha County works with the USDA Farm Service Agency to conduct damage assessments and aid in emergency agricultural assistance applications. Waukesha County businesses and residents affected by flash floods and flooding on July 11, 2017, can currently apply for low-interest disaster loans from the U.S. Small Business Administration.

USDA's Noninsured Crop Disaster Assistance Plan (NAP) provides financial assistance to producers of non-insurable crops when low yields, loss of inventory, or prevented planting occur due to natural disasters.

In Texas, over one million people were displaced by Hurricane Harvey in August 2017. Livestock were stranded after 20 trillion gallons of water (enough to supply New York City's water needs for over fifty years) fell on the Houston area in less than a week. The State of Texas Agriculture Relief (STAR) Fund, created solely with donations from private individuals and companies, assists farmers and ranchers in rebuilding fences, restoring operations and paying for other agricultural disaster relief.

"Depending on the situation, there are a lot of emergency assistance programs," Marinelli said. "Becoming familiar ahead of time, talking with neighbors, talking with other farmers to find out what they've done in tough times-- when farmers are families, and it's been generations farming, I think those lessons get passed down." New farmers can develop strategies and learn about resources by speaking with other farmers in their community.

Crafton found advice and moral support by reaching out to other farmers on social media. "Everybody had something to say. In some small way it helped to know other people were going through it too," he shared.

Two years after Orangeburg County, South Carolina's drought and deluge, Crafton reported his operation was back on track, thanks mostly to good weather. By mid-2017, Crafton had grown the operation. However, at the time of this writing, he still couldn't find insurance to cover his operation. (Perhaps USDA's NAS plan will be a viable option for him.) Yet Crafton was optimistic. "The bigger the operation, you're better able to take those hits," he said. "Another summer drought with a flood like that would still be devastating, no doubt. But I'm a little better experienced, and a little better prepared now having been through it."

More Information:
The economic impact of a disaster cannot be ignored. Read USDA's NAP fact sheet at https://www.fsa.usda.gov/Assets/USDA-FSA-Public/usdafiles/FactSheets/2016/NAP_for_2015_and_Subsequent_Years.pdf

Farm Emergency Preparedness Plan*

before a disaster or emergency
  Gather information
  - What disasters or hazards are most likely in your community? For your farm?
  - How would you be warned?
  - How should you prepare for each?

  Draw a farm site map and indicate:
  - Buildings and structures
  - Access routes (roads, lanes)
  - Barriers (fences, gates)
  - Locations of livestock
  - Locations of all hazardous substances
  - Electrical shutoff locations, etc.

  Put together an emergency supply kit for your family
  - Visit https://www.ready.gov/make-a-plan
  - Out-of-state contact person
  - List of contact #s for neighbors
  - Identify meeting place for family
  - Pet information

  Make a list of your farm inventory, include:
  - Livestock (species, #of animals)
  - Crops (acres, types)
  - Machinery & equipment (make, model)
  - Hazardous substances (pesticides, fertilizers, fuels, medicines, other chemicals)

  Keep a list of emergency phone numbers

- Your local & state veterinarian
- County extension service
- Local fire, police, ambulance
- Insurance agent

Make a list of suppliers or businesses providing services to your farm
- Livestock or milk transport
- Feed delivery
- Fuel delivery

Identify areas (higher elevation) to relocate your assets, if needed
- Livestock and horses
- Equipment
- Feed, grain, hay
- Agrochemicals (pesticides, herbicides)

Stockpile supplies needed to protect the farm
- Sandbags & plastic sheeting, in case of flood
- Wire & rope to secure objects
- Lumber & plywood to protect windows
- Extra fuel for tractors & vehicles
- Hand tools for preparation & recovery
- Fire extinguishers in all barns & vehicles
- A safe supply of food to feed livestock
- Gas powered generator

Prepare farm employees
- Keep them informed of the farm's emergency response plan
- Identify shelter-in-place or evacuation locations
- Establish a phone tree with contact information for all employees
- Know the warning signals for your area
- Learn the warning systems for your community

- Are you able to hear or see the appropriate warning from your farm?

Stay alert for emergency broadcasts
- Emergency Alert System broadcasts on radio or television
- NOAA weather radio alerts
- News sources – radio, television, internet
- Charge all electronic equipment (cell phones, tablets, laptop computers)

*Checklist provided by Vivian Marinelli, Psy.D., Senior Director, Crisis Management Services, FEI Behavioral Health

PART IV

SOIL

THE CLIMATE BELOW YOUR FEET

"In wetter periods, healthy soils act as a reservoir to infiltrate and percolate water, reduce erosion, and sedimentation; in periods of high heat, healthy soils hold available water and regulate soil temperature."

~ *Marjorie Kaplan, Associate Director of the Rutgers Climate Institute*

From carbon sink to emitter of nitrous oxide, the ground on which you stand has more power than you think. The USDA Natural Resources Conservation Service declared soil health the number one concern in 2015. The United Nations Food & Agriculture Organization (FAO) declared 2015, "International Year Of Soils." Why all the fuss? In years past, scientists viewed soil as an inert substrate that had no significant biological properties of its own. From the 1930s until recently, common agricultural practice held that this substrate could sprout life, provided it had the right inputs. After the Dust Bowl devastated the Midwestern area of the U.S. known as the "bread basket," the Soil Conservation Service was established to mitigate the effects of the dust bowl and prevent a similar event from happening again. Important land management techniques introduced in the Dust Bowl era designed to stem erosion, such as contour farming and crop rotation, fell out of favor as market demand grew. Monoculture farming (aka Monocropping) and the intensive use of chemical fertilizers and pesticides put more pressure on the soil, and in particular placed more stress on the biological components of the soil. Although it took almost half a century for the scientific and agricultural community to recognize the damage done to soil health, and another two decades to spread the word, we are beginning to understand that soil is not a lifeless substrate.

8.
PROFILE OF A MISUNDERSTOOD SUBSTRATE

With the Connecticut River meandering through it, the Pioneer Valley of Massachusetts is prized for its rich soils. Here, CSAs abound, and local farmers sell their goods at farm stands, vibrant farmer's markets, and to local grocers and national chains. Tomatoes of all kinds are a summer staple, and rich varieties of squash decorate many a table from August through December. Yet all is not perfect in this idyllic New England community. "Agriculture is becoming a victim of its success," said Massachusetts' NRCS state conservationist Christine Clarke. "There is incredible pressure on farmers to find land to produce the crops demanded by the market."

As demands for local food, plus an outbreak of Phytophthora capsici in 2015 forced some farmers to move to steep hay fields to produce their crops, Clarke grew concerned about the environmental impact. "When farmers go to fresh soil, they're often plowing down hay fields, which are a carbon sink for the reduction of greenhouse gasses," she lamented.

Phytophthora capsici, a highly infectious water mold that is lethal to certain vegetable crops including squash, pumpkins, cucumbers, potatoes, tomatoes, and peppers, infested one field after another. Crops were destroyed, and farmers watched profits dissolve. As if that wasn't bad enough, farmers suddenly had the added expense of renting new crop land that had not been infected by the mold.

"The destruction is pretty easy to see," reported Tom Akin, the NRCS-MA State Resource Conservationist who responded to the farmers in crisis. "The trail of dead plants usually follows the flow path of moisture in the soil. Where soil ponds, the impact is the worst."

One way to mitigate the effects of Phytophthora *capsici* is to improve soil health. Akin said anything a farmer does to reduce soil compaction and improve drainage can help. Could the outbreak actually have been prevented with better soil management practices?

Much Ado About Soil

Healthy soil is a living organism, teeming with bacteria, fungi, and microorganisms that must be nourished to provide a healthy environment for plants to grow and thrive. Today, the USDA's Natural Resources Conservation Service (NRCS, formerly known as the Soil Conservation Service) is promoting soil health as a means to foster a thriving agro-economy.

"We're looking at soil as a living ecosystem," explained Jimmy Bramblett, former WI state conservationist with USDA- NRCS, "and trying to maximize the chemical, physical, and biological properties of any type of soil out there."

Most current nutrient management programs focus on the chemistry of that equation. Bramblett said adding the biology and the structure to the equation enables farmers to enhance the soil and make it more resilient.

Resilient Soil, Healthy Soil

Engaging in practices that promote soil health benefits growers by saving on labor, fertilizer, water, and pesticides. Creating healthy, resilient soil is simple in some regards, yet complicated by the diversity of ecosystems in every region. California alone has over 2500 different types of soils, and more than 300 crops. In Wisconsin's soils, increasing organic matter by just 1 percent results in increased water holding capacity on the average soil in WI by about 27,000 gallons. That's great for Wisconsin, but a 1 percent increase in organic matter may not be enough in other growing areas. Idaho has two very distinct climate areas. The Snake River Plain receives 12" or less of rainfall per year, while areas of southern Idaho are basically temperate rainforest, and Florida is warm and wet. How do you know what your soil needs?

California's state conservationist Carlos Suarez depends on the information from soil surveys NRCS scientists conduct in order to help producers. "Given the producer's type of soil, and the desired crop, what type of practices do they need to use to grow that crop?" Suarez asks. "Soil health is so critical. Healthy soil retains moisture, which enables crops to grow."

Healthy soil also promotes plant health by, among other things, preventing plant pathogens and diseases from taking hold. Could the outbreak of Phytophthora capsici in Massachusetts' Pioneer Valley have been prevented? According to Akin, the disease is difficult to manage, but it's also preventable. The former agronomist said that's where the importance of managing for soil health comes in.

What's In Healthy Soil?

According to "Soil Carbon Restoration," a white paper published by the Massachusetts chapter of the Northeast Organic Farming Association (available online at www.nofamass.org/carbon), a teaspoon of healthy soil contains more microbes than there are people on earth. NRCS reported that a teaspoon of productive soil generally contains between 100 million and 1 billion bacteria, as much mass as two cows per acre. Activity of soil organisms follows seasonal as well as daily cycles. Availability of "food" is an important factor that influences the population and level of activity of soil organisms. What do soil organisms consume? Among other things, soil organisms break down carbon and nitrogen, two of the most detrimental greenhouse gases. That's the good news, because it means that bringing your soil to optimal health is not only good for your plants, and thus your operation, but also it helps to mitigate, and even slow global warming, which means you'll have more predictable weather patterns to plan around.

Microbes, Bacteria & Fungi - Oh, My!

The words bacteria and fungi may send shivers down your spine. However, only one of four functional groups of bacteria, and one of three functional groups of fungi pose a threat to your crops: the pathogens. The other three types of bacteria are beneficial: Decomposers, Mutualists, and Lithotrophs (or Chemoautotrophs) each have their function on your farm. As Ann Kennedy, of USDA Agricultural Research Service shared in "Bug Biography: Bacteria That Promote Plant Growth," certain strains of the soil bacteria Pseudomonas fluorescens have anti-fungal activity that inhibits some plant pathogens. Decomposer bacteria consume simple carbon compounds, such as root exudates, and convert energy from soil organic matter into forms useful to the rest of the organisms in the soil food web. Decomposers are especially important in immobilizing, or retaining, nutrients in their cells, thus preventing the loss of nutrients, such as nitrogen, from the rooting zone.

The Decomposer and Mutualist fungi do their part to promote soil health as well. Elaine R. Ingham reported in, "The Living Soil: Fungi," that fungi perform important services related to water dynamics, nutrient cycling, and disease suppression. Decomposers convert hard-to-digest organic material into forms that other organisms can use. Fungal hyphae physically bind soil particles together, creating stable aggregates that help increase water infiltration and soil water holding capacity.

Soil Health in Practical Terms

If the science sounds complicated, even irrelevant, it's not. Julie Rawson, of Many Hands Organic Farm in Barre, MA, altered her farming practices in 2011 when she learned about soil carbon sequestration. "Now, if I get 4.5 inches of rain - no big deal," she said. "It all goes right into the ground. And if there's a drought - no big deal. The soil holds on to that water, so the plants have it when they need it."

Weathering climate disruption and thriving in her evolving ecosystem have enabled Rawson to keep her business strong. More than that, Rawson and her team, including husband Jack Kittredge, policy director of NOFA/Mass, and author of the white paper, "Soil Carbon Restoration: Can Biology do the Job?" have been doing their part to stem the tide of climate disruption, so their children and grandchildren will be able to thrive as farmers in the coming decades.

Kittredge wrote, "Most analysts believe we must stop burning fossil fuels to prevent further increases in atmospheric carbon, and find ways to remove carbon already in the air if we want to lessen further weather crises... There is only one practical approach -- to put it back where it belongs, in the soil. Fortunately, this is not an expensive process. But it does take large numbers of people agreeing to take part."

9.
SOIL SOLUTIONS: COVER CROPS

"Figure out what your vulnerability is, what exposures you're likely to experience in your location, what are the sensitivities in your cropping system... and figure out what [you] can do about it."
~ *Laura Lengnick, author Resilient Agriculture.*

Vulnerability, exposure, and sensitivity may not be terms you thought you'd use when you got into farming. Yet as author-scientist-farmer Laura Lengnick points out, they are key to the new language of growing produce in our changing climate. Multiple options exist to determine your land's vulnerabilities, exposures and sensitivities. Perhaps the easiest place to start is with soil testing. USDA-NRCS offers free soil testing and works with farmers to develop land management plans unique to their location and cropping system.

Standard practices that benefit soil health in any location are cover cropping, encouraging friendly bacteria and fungi, eliminating nitrogen fertilizers and synthetic inputs, and direct seeding.

"A cover crop is your aspirin a day," said Christine Clarke, Massachusetts state conservationist, USDA- NRCS. "Make sure your soil is covered, and you're protecting it like your brand new toy in your garage. It's a precious resource, and a fundamental resource. If you protect your soil, you're going to inherently protect your water. It's the ultimate filter."

At Many Hands Organic Farm in Barre, Massachusetts, Julie Rawson planted clover and daikon radish under and between each row of vegetable crop. "You have to get used to a mess," Rawson explained with a smile. Cover crops create soft edges where there used to be precise rows. Beneath the "mess," a web of activity is happening that boosts soil health and mitigates climate disruption.

The daikon radishes' deep root system encourages earthworm activity and more microbial activity than shallow root crops. It increases the amount of water infiltration in the soil profile. In general, deeper root crops create a better soil health profile than shallow root crops, like grasses and forbes. As water travels deeper into the soil it reduces compaction issues. "If you have a clay pan a foot or so down because you've plowed the land for years and years, these roots are strong enough to penetrate the clay pan, break that up and provide a deeper soil profile," explained Curtis Elke, Idaho state conservationist, USDA-NRCS.

Cover crops with deep roots help mitigate climate disruption by pulling more carbon from the air and then releasing more liquid carbon into the soil than shallow roots. That liquid carbon feeds the microbes that create soil aggregates. Aggregates produce more porous soil, which holds water better in droughts and in floods. In some areas heavy rain can delay planting and create problems obtaining a good stand of plants, which can reduce crop productivity. As reported in the 3rd National Climate Assessment, in soils with even modest slopes, rainfall of more than 1.25 inches in a single day leads to runoff that causes soil erosion and loss of nutrients and can lead to flooding.

"If we want to survive we really have no alternative but to restore carbon to the soil," Jack Kittredge reported in Soil Carbon Restoration: Can Biology Do The Job? "That this can be done through biology, using a method that has worked for millions of years, is exciting... Farmers can follow these simple principles and not only restore carbon to the soil but help rebuild the marvelous system that nature has put in place to renew our atmosphere while providing food, beauty and health for all creation."

Integrating Cover Crops Into Your Cropping System

According to Idaho's state conservationist, there's not one cover crop mixture that works for everybody or works throughout the year. Still, he encourages many farmers to rotate cash crops on a four-year cycle. Alternating potatoes with wheat, sugar beets, and then another grain is a four-year rotation mixing high-residue with low-residue crops. Although wheat provides the soil significant straw residue, Elke said the common practice of rotating only potatoes & wheat is not enough. "That's why we're encouraging cover crops in low-residue crop years," he explained. "Right after the potato or beet harvest, you go in and plant that cover crop to add that extra residue and organic matter into that soil profile."

Florida's state conservationist Russell Morgan said traditional winter cover crops, such as cereal rye, crimson clover, vetch, and brassicas work well with northern Florida's row crop production systems. For the southern Florida vegetable production areas, he suggested summer growing cover crops, including 'exotic' legumes such as hairy indigo, sunnhemp, cowpeas, or velvet beans, in combination with sorghum-sudangrass or pearl millet.

According to Morgan, Florida's diverse agriculture presents many challenges to soil health. Crops include vegetables like potatoes; a myriad of specialty crops such as citrus, blueberries, and sugar cane; and row crops such as corn, cotton, soybeans, and peanuts. In addition, high year-round temperatures and rainfall make it harder to increase the organic matter content in Florida soils. "Luckily, Florida has more options for utilizing more and different cover crops than many areas of the country because we have both winter and summer production systems," said Morgan.

In Massachusetts, conservationists reported shifts in pest activity - including different timing and locations - resulting from climate disruption. The solution to such shifts is not newer, stronger pesticides, but rather soil health. "The soil biota is no different than the biota in your gut," Clarke explained. "Everything has to be balanced to work correctly." Just as friendly bacteria in the human digestive tract encourages physical health in part by discouraging unfriendly bacteria, so a healthy soil biota encourages plant health in part by discouraging pests and encouraging beneficial insects.

"Farmers cannot do much to change their exposures, but they can do a lot to change their sensitivity and their adaptive capacity," Lengnick said. In conjunction with soil testing, the framework in Lengnick's book, Resilient Agriculture, may be a useful tool to show you the areas of vulnerability in your growing operation over which you have control.

10.

NO-TILL FOR SOIL HEALTH?

"Rebuilding soil health is one of the most important steps producers can take to protect their operations from the challenges presented by the extreme weather events that climate change exacerbates."
~ Jean Steiner, USDA Supervisory Soil Scientist, Grazinglands Research Laboratory

Like an intricate puzzle, soil health has so many different pieces it can be challenging to see what they are and how they fit together. One piece in rebuilding soil health is no-till farming. Also known as "direct seeding" in some circles, no-till farming involves keeping soil relatively undisturbed and protected with residue leftover from cover crops. "No-tillage management is a key component in improving the health of the soil," said Jean Steiner, of USDA-ARS Grazinglands Research Laboratory in El Reno, Oklahoma.

The problems with tilling can be summed up in two words: disturbance and compaction. Disturbing the soil releases carbon dioxide into the atmosphere, which contributes to the rise in global temperature. Soil compaction occurs on the surface whenever a vehicle drives across a field. Plowing and disk harrowing cause compaction lower in the soil profile by "smearing" the soil at the bottom of the tillage implement. After multiple years of tillage, a thin, dense layer of soil known as a "plowpan" develops that neither roots nor water can penetrate.

There are numerous benefits to no-till farming. However, some scientists fear no-till is being promoted as a silver bullet to mitigate climate disruption. In fact, no-till can be misapplied, and have no positive climatic effect.

Matthew Ryan, of Cornell's School of Integrative Plant Sciences warned that in certain soil types and environmental conditions no-till can actually increase soil nitrous oxide emissions. For example, to offset productivity decreases that can occur with the transition to no-till management, one may be tempted to increase use of nitrogen fertilizer, which is made using large amounts of natural gas. The added nitrogen use cancels out fossil fuel reductions associated with on-farm fuel use.

When properly applied, no-till management presents no disadvantages to climate health, according to Jerry Hatfield, co-author of the 3rd National Climate Assessment. "Keeping the soil in place and preventing erosion is necessary to offset the more extreme precipitation events, which outweighs other potential detriments."

Benefits of No-Till for Mid-to-Large Scale Growers

No-till and the associated maintenance of year round vegetative soil cover offer many operational benefits:

- Protection from erosion and soil crusting,
- Increased infiltration of rainfall,
- Moderation of soil temperature,
- Reducing evaporation rate to maintain a moister soil, and
- Improved soil structure that allows for more timely agronomic operations.

In addition, adoption of no-till mitigates greenhouse gas emissions through reduced fuel usage. No-till also increases water and nutrient holding capacity, increases soil biology, and allows for better nutrient cycling. Financial benefits of no-till management include reduced input costs. It requires less fuel, and over time with an improved soil biology and nutrient cycling - less fertilizer. Reduced labor cost result from fewer field operations and the ability to farm more area with less labor. Stability of crop production increases, as untilled soils are more resilient to weather variation. Overall land value also increases, as it becomes more efficient in crop production.

Is No-Till Management Right for Your Growing Operation?

Exploring options with guidance from your local ag-extension service or USDA-NRCS agent will help you determine whether no-till management is right for you. Issues to consider include crop choice, growing region, and adoption costs.

No-tillage management does not work well with all crops. As Tom Akin, a state resource conservationist with NRCS in Massachusetts explained, no-till is most successful with large-seeded crops, which store more energy than smaller seeds. Greater seed energy leads to greater chance of developing into a healthy seedling. Germinating seeds need to emerge from the soil, and seedlings need to develop leaves to capture solar energy. No-till soils can have varying amounts of plant residue that are not hospitable to the survival of small seeds, like carrots. Using pelleted seed (seed encased in an inert material to provide protection), is one adaptation that may allow growers of small seeded crops to successfully adopt no-till management. Another adaptation - using trash whippers move residue away from the seed zone, and planting into the narrow area ensures good seed-soil contact.

Ryan said no-till works well for farmers in warm and dry environments, especially if they do not use irrigation. "If no-till is not feasible, farmers should consider conservation or reduced tillage practices, such as strip till, which can work well even in cooler environments." Yet some farmers as far north as Saskatchewan, Canada have succeeded with no-till, according to Idaho state conservationist Curtis Elke.

As with any new farming method, you may experience a steep learning curve. You may require specialized no-till planters, no-till drills, or other equipment designed to make no-till feasible in a larger growing operation. No-till planters differ from conventional planters in two major ways. First, a no-till planter usually has a leading "coulter" or interlocking finger-like row cleaners that cut through plant residue or "clean"/move the residue out of the seed row. No-till coulters can usually be added onto conventional planters. Second, no-till planters' heavier implements have down-pressure springs that keep the planting units at a uniform depth relative to the heavy frame of the planter. Most conventional planters travel freely across the soil, with the planting units rising and falling with the micro-topography of the soil's surface.

Return on investment in no-till may be delayed by a few factors. Purchasing new equipment increases start-up costs. If soil is extremely diminished, it may take a few years for the soil biology to recover and respond to no-till. Adopting a new way of management means learning how the system responds to your unique operation. This can take time and there may be some temporary loss of productivity associated with the learning curve. Changing to no-tillage may also require a change in the nutrient or pest management program.

Attending field days, talking to other producers who are succeeding at the systems being considered, and seeking support from Extension and USDA Conservation professionals can help you succeed.

Organic No-till Vs. Conventional No-till: The Climate Challenge

If conventional No-Till can be effective, can Organic No-Till be even more effective in lowering atmospheric carbon dioxide? Can yields achieved with organic no-till agriculture match yields achieved with conventional no-till?

Like so many experiments, the early days of Organic No-Till farming, aka. cover crop-based organic rotational reduced tillage (CCRRT), research were fraught with challenges. Mark Schonbeck, of Virginia Biological Farming, reported in eXtension.org that results were inconsistent, and both weed control and vegetable yields sometimes fell short of then current standards. Schonbeck found that planting vegetables through mulch could delay vegetable growth and maturity and promote problems with slugs, cutworms and crop diseases. The randomly-oriented cover crop residues left by mowers, scythes, and other manual cutting tools interfered with mechanical no-till transplanting. Most standard vegetable planters and transplanters do not function well in untilled soil.

In recent years, CCRRT has gained positive momentum and researchers throughout the U.S. reported that Organic No-Till is feasible for vegetable growers. The fact that CCRRT is also beneficial for the climate is a boon for growers and eaters worldwide.

"A large part of the carbon footprint of conventional systems is the manufacture of the synthetic pesticides and fertilizers, which are absent (or very, very limited) in organic," said University of Wisconsin's Erin Silva. Synthetic pesticides and fertilizers are derived from fossil fuels. Nitrogen fertilizer releases nitrous oxide (N_2O) into the atmosphere.

The same inputs that reduce the need for synthetic nitrogen fertilizer also sequester more carbon from the air and put it back into the soil where it belongs. "We don't have any other greenhouse gas data yet," Kathleen Delate of Iowa State University reported, "but by putting more carbon in the ground, you can help mitigate GHG emissions."

Is CCRRT/ Organic No-Till Feasible for Your Farm?

By protecting soil from erosion while increasing both water infiltration and soil moisture holding capacity, CCRRT helps you deal with flooding and drought, thus enabling you to weather at least one symptom of climate change. "This past year we observed greater water infiltration in rolled-crimped cereal rye compared to soil without mulch. This can help the soil absorb water during intensive rainfall events, which have increased over the past 50 years," Cornell University's Matt Ryan shared. "We observed greater soil moisture late in the summer during a short-term drought period in organic no-till plots compared to plots managed using traditional organic methods. This means crops will have access to more stored water if conditions become dry."

"Organic no-till is still in its infancy and the best results have been with annual crops that have large seed sizes such as soybean and corn," Ryan said. CCRRT is fairly scale-neutral and can be practiced on large farms.

More work is needed to determine how CCRRT might best be implemented on vegetable operations. At Virginia Tech, Ron Morse did copious research on organic no-till vegetable production. Experiments seeding pumpkins into cereal rye produced superior quality pumpkins, due to the fact that the fruit was not sitting on bare soil.

At University of Wisconsin, results of research with vegetables varied. Silva found CCRRT works best for transplanted vegetables, and she stressed the importance of delivering adequate fertility to the vegetable crops at the appropriate time. Specialized transplant equipment may be needed. "In some cases, mulch helps with fruit quality and suppresses pathogen spread - in other cases, it may exacerbate it," Silva explained. "Strip till may be a more feasible approach, with the no-till phases being limited to the alley ways - this may lead to easier management of the vegetable crop, including the use of other weed suppressive mulches."

Experiments with corn-soybean-wheat rotations show organic farmers can reduce on-farm fuel use and labor by approximately 25-33 percent when transitioning from tillage based organic crop production to CCRRT production. There is currently no data for reduction in fuel use and labor in organic no-till vegetable operations.

Practical Differences Between No-Till and Organic No-Till

Inputs: Conventional no-till relies on the use of synthetic herbicides to manage weeds and nitrogen fertilizer to boost yields. The critical input in organic no-till is the cover crop, which creates a thick mulch that suppresses weeds during the cash crop production season. As this is the primary input, management of the cover crop is essential. All cover crop management should focus on creating the amount of cover crop biomass required for reliable weed suppression - around 8,000 - 10,000 pounds dry matter of cover crop per acre. Instead of nitrogen fertilizer, legume cover crops provide nitrogen for crops planted using CCRRT method.

Tools: Planting equipment is similar in both conventional and organic no-till systems. The tools for killing cover crops differ. CCRRT systems use a roller-crimper to kill cover crops, which serve as mulch and effectively replace the herbicides commonly used in conventional no-till production.

Many different styles of roller-crimpers exist, but Silva reported good results with the Rodale-designed model. Front mounting the roller crimper on a three-point hitch and planting in a one-pass operation offers consistent termination of the cover crop and savings of fuel and labor. "Emerging [cash crop] seedlings must be able to penetrate through the mulch, which lends itself to larger seeded crops with more stored energy," she explained.

Methodology & Timing: Planting later in the spring allows cover crops in an organic no-till system to reach the reproductive growth stage, so they can be effectively terminated without herbicides. Delate found the conventional practice called "continuous no-till" impractical in dealing with perennial weeds that can build up. "We have gone to tilling every third year of the rotation. You will still get the benefits of organic no-till every season you don't till," Delate said.

Direct seeding, or drilling directly into the dead cover crop or stubble of previous crop, can be done for organic no-till but is more typical for conventional, since growers often plant on 30-inch rows and don't use a drill. Delate and her colleagues are experimenting with the difference this year.

ROI

It's all fun and games until you see a return on your investment. Your economic success will depend on many factors. If you already have no-till planting equipment, the only additional cost is for a roller crimper, which can cost approximately $3500 for a 10 ft unit (http://www.croproller.com/). Returns will depend on the weather each year, but should be greater than returns from traditional management practices in dryland farming in years with either too much or not enough rain.

Which system offers the best solution for mitigating global climate disruption? Delate, Ryan, Silva, and others seem to agree CCRRT is the winner. This production method sequesters more carbon and emits fewer greenhouse gases, but Delate acknowledged it is a complicated system that may not work every year. "We are still tweaking the system," she said.

Read more about No-Till at http://articles.extension.org/pages/28317/reducing-tillage-to-save-fuel

PART V

BEYOND CARBON

"There are big opportunities and challenges for sustainable food production within a changing climate. Climate change is connected to many parts of the food system that need to be considered in concert: yield trends, water availability, water quality, habitat loss, diet, and waste. Each of these factors influences the impacts of climate change on the food system."

Paul West, Co-Director and Lead Scientist at University of Minnesota's Global Landscapes Initiative.

What role DOES agriculture play in exacerbating climate change? Agriculture is one of the biggest sectors contributing to total greenhouse gas emissions — currently about 20-25 percent of total global emissions. Nitrous oxide accounts for about 6 percent of all U.S. greenhouse gas emissions. The gas is naturally present in the atmosphere as part of the Earth's nitrogen cycle, and has a variety of natural sources. However, the EPA reports human activities are increasing the amount of N_2O in the atmosphere. Agricultural soil management is the largest source of N_2O emissions in the U.S., accounting for about 75 percent of total U.S. N_2O emissions in 2015. Nitrous oxide is also emitted during the breakdown of nitrogen in livestock manure and urine, which contributed to 5 percent of N_2O emissions in 2013.

Nitrous oxide emissions from agricultural soils were about 18 percent higher in 2013 than in 1990, and emissions were projected to increase by 5 percent between 2005 and 2020, driven largely by increases in emissions from agricultural activities.

Of course, you have the power to curb this. Agriculture has the chance to play a critical role in national strategies to reduce greenhouse gas emissions. According to West, the agriculture community's opportunity is also its responsibility.

11.
FARMING'S ROLE IN THE NITROGEN CYCLE

"If you follow the nitrogen you'll find the hotspots," said Michigan State University's distinguished professor of ecosystem science Phil Robertson. He was referring to the concentration of nitrous oxide emissions found near farms with fertilizer-intensive cropping and intensive animal production.

The application of nitrogen fertilizer, whether as a synthetic input, manure, or organic leguminous cover crop, is so common in agriculture that many growers believe it is an absolute necessity. Some farmers disagree. Jack Kittredge, past president of the Massachusetts chapter of the Northeast Organic Farming Association, said bacteria sequester nitrogen from the air for plants to absorb - provided the soil is healthy. This process, called nitrogen fixing, has enabled agriculture for millennia. "Anyone who thinks synthetic nitrogen fertilizer is necessary is not dealing with healthy soil," he shared. "It is the use of synthetic chemicals which has destroyed much of the biology in the soil. The result is lifeless soils that can no longer perform vital functions."

The necessity of nitrogen fertilizer isn't just an esoteric debate; it's critical to sort out in our changing climate. While media widely report the warming effects of carbon dioxide on the atmosphere, few outlets mention that the impact of 1 pound of nitrous oxide (N^2O) on warming the atmosphere is almost 300 times that of 1 pound of carbon dioxide (CO^2). Nitrous oxide molecules stay in the atmosphere for an average of 114 years before being removed by a sink or destroyed through chemical reactions. Further, the microbes that convert soil nitrogen to nitrous oxide "don't much care where their nitrogen comes from," Robertson explained. "So pound for pound organic fertilizers like manure result in about as much nitrous oxide emission as synthetic nitrogen." The same appears to be true for the nitrogen in legumes, although cover crops may prevent some nitrous oxide emission in the fall through spring period by capturing soil nitrate that would otherwise be emitted as nitrous oxide. Experimental work on this point is scant, Robertson noted. "The basic general fact is that the more inorganic nitrogen there is in soil the more nitrous oxide will be emitted."

Changing Soil Management Practices Can Lower Nitrogen Emissions

Adopting management practices that foster soil health can help to keep nitrous oxide out of the atmosphere. In her report The Living Soil: Bacteria for USDA's National Resources Conservation Service, Elaine R. Ingham described one aspect of the science behind healthy soil. "Mutualists bacteria form partnerships with plants. The most well-known of these are the nitrogen-fixing bacteria... Some of these species are important to nitrogen cycling and degradation of pollutants."

Bacteria alter the soil environment such that it will favor certain plant communities over others. Before plants can become established on rocks and other geological surfaces, the bacterial community must establish, starting with photosynthetic bacteria. These fix atmospheric nitrogen and carbon, produce organic matter, and immobilize enough nitrogen and other nutrients to initiate nitrogen cycling processes in the young soil. Then, early successional plant species can grow. As the plant community is established, different types of organic matter enter the soil and change the type of food available to bacteria. In turn, the altered bacterial community changes soil structure and the environment for plants.

Removing nitrous oxide from the atmosphere is a double-edged sword. The gas is naturally removed from the atmosphere mainly by chemical reactions in the stratosphere, but that also depletes stratospheric ozone, contributing to the infamous ozone hole over polar regions. While CO_2 is removed by photosynthesis and chemical reactions in the ocean, Robertson explained, "The better strategy for reducing nitrous oxide is to avoid excess emissions altogether rather than try to remove what's there."

Multiple studies have revealed nitrogen applied in fertilizers and manures is not always used efficiently by crops. Typically only half of the nitrogen fertilizer applied is actually taken up by the crop during that growing season, according to a Michigan State University extension fact sheet. Improving efficiency can reduce emissions of N_2O generated by soil microbes, largely from surplus nitrogen.

You can reduce nitrous oxide emissions from your farm, while maintaining yields and crop quality, by using the 4 R's for better nitrogen fertilizer management. Right rate means applying no more than the crop can use. Right time means applying as close to when the plant needs it as possible. Right place means applying as close to plant roots as possible. Right formulation means applying forms of fertilizer that are most likely to stay in the soil rather than gassify.

Fertilizer added in drip irrigation systems results in far fewer nitrous oxide emissions than does fertilizer that is added indiscriminately.

"Conveniently, management that leads to avoided nitrous oxide emissions also leads to fewer losses of nitrate and other forms of reactive nitrogen," said Robertson. Thus there are co-benefits to better managing nitrogen, including economic savings when less fertilizer is used more precisely. Likewise cover crops provide co-benefits by building soil carbon and reducing erosion. No-till can do this as well.

12.

NEW TECH IN N²O SAMPLING

Carbon emissions get a lot of airtime in the media, but increasingly there's a consensus understanding that nitrous oxide (N²O) is the most important greenhouse gas for Ag to address. At North Carolina State University (NCSU), one group of researchers is focusing on nitrous oxide, in part because emissions data coming from the southeastern US is very different from data coming out of other parts of the country. Researchers have been able to recognize this difference in recent years thanks to a new measurement system designed by Professor Wayne Robarge, a soil scientist at NCSU's College of Agriculture and Life Science.

"What happens for us is we're emitting almost none and then we get a big rain storm that really soaks the soil profile and we'll have a brief burst [of N²O emissions] that's pretty big and then dies off after a couple days," said Chris Reberg-Horton, Associate Professor and Organic Cropping Specialist at NCSU. "Identifying that it's episodic and figuring out how to measure that has really been a big deal. We've spent a lot of effort on that. Now, we have a system that monitors twenty-four hours a day. Folks in other places have been taking, like, weekly measurements of N²O. We have a system that watches the storms come through and monitor the N²O with each storm. Now we're ready to go back and re-do some of our comparisons."

Reberg-Horton said the data raises the possibility that nitrous oxide emissions in organic agriculture and conventional agriculture are driven by different things. The timing of when the greenhouse gas comes out of the organic system and conventional system is different enough that Reberg-Horton, Robarge, and their colleague Shuijin Hu are wondering if one system is better than the other in this regard. "We don't have enough info yet to determine that," Reberg-Horton said.

In conventional farming, N_2O emissions are driven by big pre-planting fertilizer bursts. At NCSU's experiment station in Goldsboro, researchers apply 75# of nitrogen before planting corn, and Reberg-Horton said that's the application that's most vulnerable to losing nitrous oxide. They also apply 75# of nitrogen as side dressing, which tends to stay put. "We think it has to do with the water," said Reberg-Horton. "When we put it out pre-plant, you've got this tiny little plant, and it's also the cooler part of the spring, so the soil can stay wet and that's when we see nitrous oxide emissions. Later in the season, when we side dress, the plants are already big and it's hard to keep the ground saturated. The big plants suck up the water so fast that we just don't see as much N_2O coming out of it."

The results are most dramatic in crops that require more nitrogen fertilizer, specifically corn. Organics look a little different, and Reberg-Horton said this is because organic farming systems don't allow large applications of nitrogen at any one time.

"We have a hypothesis, but we don't have the data to support it yet," said Reberg-Horton. "We're still working on the data. In conventional systems we have so much excess nitrogen available that in the southeast what regulates it is how much carbon is in the soil. The limiting factor is how much carbon is in the soil. In organic systems, the limiting factor is the amount of free nitrogen. The way each system fundamentally regulates how much N_2O is emitted is different. It still leads to this question, which is frustrating: which system emits more, and less, as you go all the year round? Or even better yet, which system emits more when you go all the way through the [three-year] rotations of corn, and soy and wheat."

Getting to Theory

To reach the point where Reberg-Horton, Robarge and their colleagues could even develop their hypothesis required a mindset shift. First, they realized they had to adjust how nitrogen emissions are typically measured. In the old-school method of sampling nitrogen emissions from soil, one uses hand syringes to pull samples from chambers laid on top of the soil. This method of research typically occurs weekly or during or soon following a rainstorm. "The when-to-sample issue was done much more cavalierly," said Reberg-Horton.

The new method is much more precise. "What we have is a trailer set up in the field, plugged into a power pole that we installed out there. We've got vacuum pumps connected to these long hoses that go out into different parts of the field. Those vacuum hoses are pulling air from different chambers scattered throughout the field. The most essential part is: the soil needs to be in the open - rained on and exposed to sun and everything else like normal soil, but then we need to periodically take gas sampling off that."

To do that, the NCSU researchers use several robotic arms, each with a lid. The researchers drive something into the soil that that each lid can sit on. Every 30 minutes, one or more of the robotic arms places its lid on the surface of a chamber, pulls gas for a few minutes, feeds the data into the field trailer where the researchers can analyze it, and then removes the lid from the soil, exposing the soil to natural conditions again. Having multiple sampling stations in the field enables the researchers to take a gas sampling from one or more places in the field all the time. They rotate which chamber in the field is feeding data into the trailer.

Soil scientist Wayne Robarge developed a technique (and the machinery required) to connect this continuous sampling method with the old-fashioned chamber sampling method. "For agricultural research we really need to do things repeatedly. We need to have replication," said Reberg-Horton. "We use [Robarge's] machine that's running all the time to tell us 'Hey there's a big gas event happening right now; it's time to sample.' As soon as the gas event is over we don't sample anymore. So it's the machine informing us when are the key times to sample that has been the big breakthrough."

As a result, Robarge, Reberg-Horton and their team now know not only exactly when to sample, but also how to plug the data into the larger set of information. "If there's an event - like a two day event that happens, if we go in and do just a couple of hand samples, we know where those samples fit in the entire pulse of events, so we can interpret the data differently."

As yet, researchers have not figured out how to apply the data, but Reberg-Horton anticipates it may lead to an alteration in fertilizer formulations. "We haven't studied that yet, but anything that reduces the amount of free N being released all at once, in theory, will help," he said. "The ideal would be that we could somehow alter our system management to help prevent [nitrous oxide emissions], and we don't have good advice on that yet. There's some very early evidence that there are other chemical compounds that are not used right now in the fertilizer industry that might inhibit N^2O emissions. That would be promising."

PART VI

WATER

To understand some of the problems farmers worldwide face with water, you first need to know how it moves underground. Water that seeps into the ground and is not taken up by plants travels deeper, through an aquifer of fractured rock, gravel and sand until it reaches a layer of rock it cannot easily penetrate. The water that pools there is called groundwater.

Groundwater constitutes much of the world's irrigation supply. What happens when there's long-term drought? If a deluge follows, does it refill the aquifers that house an area's groundwater? After four years of drought, California was finally starting to see an increase in precipitation in 2017. Despite the increase, the U.S. Drought Monitor reported in June 2018 that more than 50 percent of California's soils remained "abnormally dry," with much of Southern California's soils in "moderate drought," to, "severe drought."

Since California farmers grow a vast majority of our nation's produce, the ongoing drought is a significant concern. Yet it's also an opportunity for California's farmers and ranchers to build resiliency for the more intense droughts expected in coming years. Groundwater resources management and quality protection can provide a sustainable future at regional, national and global scales. In these pages, a few ideas you can apply to manage your water resources, no matter where you're farming.

13.

WATER
Adapting When There's Too Much or Not Enough

On a mountainside in Downeast Maine, Gail VanWart and her husband turned their 150-year-old wild blueberry farm into a native pollinator sanctuary. The move, a creative attempt to save the blueberry crop, was a response to several years of intense springtime rainfall that was preventing imported honeybees from pollinating the plants. In Hammonton, New Jersey, Bobby Galletta, of Atlantic Blueberry Company installed a network of drainage to handle floods, plus overhead drip irrigation for the dry seasons. In other parts of the country, fruit growers have attempted to prevent cracking or spotting from too much moisture by using helicopters over their fields to dry fruit.

    Meanwhile, vegetable growers like Jim Crawford of Hustontown, PA adapted to extreme wet weather by amending soil, shaping beds in fall and winter, and protecting beds with black plastic mulch so they can plant on time in spring, no matter the weather.

    Dwindling water supplies in Center, Colorado motivated potato grower Brendon Rockey to experiment with cover crops to reduce water use. The new, more diverse cropping system maintained yields and crop quality while using 50 percent less water and reducing the need for fertilizers and pesticides. Improved soil quality was an unexpected side benefit for Rockey's potato crop. Promoting soil health is the key mitigation strategy currently in place for both drought and flooding at The Farm School, a working farm and educational facility in Athol, Massachusetts.

To offset increased precipitation in Iowa, producers have been installing subsurface drainage to remove water from the fields - at a cost of $500 per acre. In central California's Pajaro Valley, farmers produce nearly $1 billion in fruit, vegetables and flowers on approximately 28,000 irrigated acres each year. A long-term, community-led effort to address irrigation issues associated with groundwater overdraft and seawater intrusion in the region led to increased use of recycled water for agriculture, capturing storm flows for later use, and water conservation efforts to reduce demand.

Strategies for adapting to climate change can sometimes be costly, but will be necessary if weather extremes and sea-level rise increase as projected. Of course, sea-level rise will affect coastal farmers worldwide. As an example, coastal flooding will reduce available farmland in Florida, a major vegetable production state. "As saltwater continues to intrude into aquifers that are being used by agriculture, there is a threat that many more farm fields will have to be abandoned due to lack of freshwater for irrigation," explained Chris Obropta, Rutgers University's Extension Specialist for Water Resources.

A Critical Resource Challenge

According to Laura Lengnick, a climate resilience planning consultant and author of Resilient Agriculture, the availability of water has been identified as one of the most critical natural resource challenges facing US agriculture in this century. Agriculture uses eighty percent of our nation's freshwater, and 90 percent of freshwater in the western states. Some agricultural practices have contributed to degradation of water quality and reduced the quantity of freshwater supplies. "Climate change will create additional challenges to water resources, because higher temperatures and drought will increase water demand, while more variable precipitation and heavy rainfall will increase soil erosion and runoff and reduce groundwater recharge," Lengnick reported.

In California, where more than 50 percent of our nation's fruits and vegetables are grown, water shortages idled more than 500,000 acres of farmland in 2015, at an approximate cost of $1.84 billion plus loss of over 10,000 seasonal agricultural jobs. It was considered to be California's worst drought in over 100 years, since people started recording weather patterns. Despite 2016 springtime rainfalls in the state, the California drought continued into 2017. Net water shortage was expected to reach around 2.9 million acre-feet per year, groundwater elevations will continue to decline (some inland farm fields are actually sinking), as will the ability to tap groundwater for irrigation (or any other use).

Decades of overuse of groundwater in central California's Pajaro Valley caused the migration of seawater into the freshwater aquifers, leading to soil degradation and decline in crop quality. Brian Lockwood, of the Pajaro Valley Water Management Agency (PVWMA), reported that increased groundwater salinity rendered the region's strawberries less productive on a per acre basis compared with strawberries grown with higher quality water.

"Every crop has a certain tolerance for salinity," explained UC Davis professor & hydrologist Helen Dahlke. "As salinity levels increase in the groundwater or irrigation water there is an increased risk for crop damage, with reduced crop quality and yield."

Strategies

As part of his mission to educate farm workers, farmers and legislators about water, Samuel Sandoval-Solis, assistant professor in UC Davis' Department of Land, Air and Water Resources and Cooperative Extension Specialist, shares information about irrigation solutions that work in climatic shifts. "It's a mosaic of different policies, from improving storage through modifying reservoir operations, to modifying conveyance of water, improving aquifer recharge and storage, [and] water conservation," he said. The policies and solutions often work together, and none is effective enough as a stand-alone solution. As Solis emphasized, "The silver bullets are gone."

Described by Solis as "water insurance," Aquifer Recharge is a natural process that occurs when precipitation or melt occurs regularly and in sufficient quantities to raise the level of the water table. Aquifer recharge can also be replicated artificially, by diverting water and putting it into the ground at targeted insertion points. Solis said the practice benefits all the farmers that withdraw water from that specific aquifer. "Water will move faster or slower depending on the aquifer, but the aquifer is connected, so ultimately everyone will benefit."

In central California's Pajaro Valley, where reduced groundwater pumping created the most significant reduction of seawater intrusion in recent years, PVWMA adopted a strategy to develop new, reliable sources of water for agricultural irrigation in lieu of groundwater production. Sources included the Water Recycling Facility, the Managed Aquifer Recharge and Recovery Facility, and a Coastal Distribution pipeline designed to convey water from these facilities to over 7,000 acres of farmland located in the middle of the seawater intrusion zone.

PVWMA has been working with growers to help improve irrigation efficiency, or, as Brian Lockwood said, "working to make every drop count." As part of that effort, PVWMA offers a rebate for enhanced recharge.

Working with upstream neighbors to recharge groundwater aquifers during wet years is a strategy also recommended by Dr. Dahlke. It allows more groundwater to be available during drought years when surface water supplies are limited.

Reducing Water Demand is a simple but challenging concept to implement. It involves identifying crops that require irrigation and steering resources toward such crops while also identifying crops that do not require irrigation and diverting resources from those crops. As reported in Southwest Farm News, Texas A&M researchers are responding to decreasing water levels in the Ogallala aquifer by exploring the viability and infrastructure required to switch from high-water-use crops like peanuts, potatoes and corn to vegetables like tomatoes, that require less water.

One of the three most important actions all produce growers can take to prepare for drought, according to Obropta, is to convert to smart water conservation irrigation systems. "Drip irrigation will be much better than other systems that use more water," he explained. Selecting the right growing location, outside of floodplain areas, is another adaptive strategy Obropta suggested.

Idaho receives only twelve inches of rainfall each year, so Idaho's USDA-NRCS pays landowners for property easements, puts unproductive agricultural land into a reserve program, and stores water on that land. "It becomes a collection basin," Idaho State Conservationist Curtis Elke explained. "…basically for water, wildlife habitat and migratory birds. It also serves as a recharge for the states's aquifers and reservoirs and other things, so it's multipurpose benefit."

As Howitt et. al reported in "Economic Analysis of the 2015 Drought for California Agriculture," new groundwater regulations designed to force groundwater basins towards sustainable yields in California may initially cause increased fallowing or longer crop rotations, but will preserve California's ability to support more profitable permanent and vegetable crops through drought.

Whether you choose high-tech or low-tech Infrastructure improvements, understanding your water resources is beneficial. In New Jersey, Galetta has experienced both drought and flooding. Thus far, he is getting by with the on-farm water supply produced by natural and shallow wells. In Maine, the VanWarts have an irrigation pond, but remain unconcerned about flooding, due to their high altitude.

At the Farm School, Neukirch and his colleagues are considering whether to use the topography of their landscape to help move and hold water on the site. "In the immediate area surrounding the farm, residential wells are often over 400 feet deep and produce a very low flow rate," Neukirch said. "This makes drilling wells for irrigation an economically dicey proposition and increases the potential viability of directing and holding surface and subsurface moisture in ponds, swales, etc to increase the resilience of our landscape. Such water retention strategies coupled with irrigation infrastructure that maximizes the utilization of drip irrigation, continued investment in increasing soil organic matter and whole farm eco-system resilience are our greatest strategies for mitigating drought, flooding, and groundwater salinization."

PVWMA built a supplemental water supply pipeline to provide Recycled Water (tertiary treated, disinfected wastewater) for irrigation as an alternative to groundwater pumping. "With the completion of the Coastal Distribution System, we've been able to provide a supplemental supply of water in-lieu of groundwater production," Lockwood reported. "The delivered water quality is better than the quality of wells suffering from seawater intrusion, and the delivered water has the added benefit of reducing groundwater production in an overdrafted basin."

As Obropta suggested, designing stormwater management practices will be beneficial during flooding. Rainwater Storage Systems can provide water during dry periods.

If you're farming traditionally, your Soil Health may actually be your most important infrastructure investment. A 1 percent increase in soil organic matter in the top six inches of soil in an acre of land can result in as much 25,000 additional gallons of available water to crops. Higher organic matter produces better infiltration capacities, conveys water faster through the soil profile, and reduces waterlogged, anaerobic conditions in the soil.

Quick Tips for Weathering Drought & Flooding

*Do not irrigate when water is not needed.
*Encourage deeper rooting of crops.
*Plant annual crops that could be fallowed in severe drought periods when no water is available.
*Plant a diverse array of crops.
*Choose a high value crop to maximize return and offset costs of irrigation, etc.
*Introduce low water use or drought tolerant varieties.
*Buy crop insurance. (note: Insurance is not available for all crops.)
*Plant on both high ground and low ground if possible.
*Have money in savings to carry you through a loss year.
*Call Extension for instructions on how to salvage crop.
*Invest in water conservation practices, like drip irrigation, so that overall water use during the high-evapotranspiration periods is reduced.
*Harvest rainwater.
*Recycle water on your farm: capture unused water from greenhouses and reuse it.
*Collaborate with area farmers, or join an existing network, like California Climate and Agriculture Network, for support and ideas.

14.

THE ART OF WATER

New Concepts in Irrigation Infrastructure

Irrigation is an ancient concept, developed 8000 years ago in Egypt and Mesopotamia and refined over millennia. Some irrigation practices have stood the test of time, while others have given way to technological advancements. As the population continues to grow, good irrigation becomes even more important. Good irrigation leads to better yields. Irrigation systems created without environmental stewardship in mind have benefited agriculture in the short term while harming fish, wildlife, and native flora species. In the 1990s, a popular argument against saving the spotted owl circulated throughout the media. The argument suggested that focusing on wildlife health was a waste of resources with no benefit to society. Science has now proven that argument to be wholly unsound. Each species is part of an intricate web of interdependence; ultimately taking care of the lowliest creatures does benefit humanity and even the business of agriculture. Nowhere has this been proved more clearly than in California and in Washington State's Yakima River Basin, where environmental degradation caused by agriculture ultimately hurt farming.

California's five-year drought ended with a deluge in 2017, which resulted in flooding and drainage issues. Yet analysts predict the long-term reliability of water supply will continue to be a major issue in California agriculture well into the future. UC Davis Research Engineer Josue Medellin said groundwater management needs to be addressed from an institutional point of view and farmers need to organize themselves, so the resource is managed sustainably to face future droughts.

California's Sustainable Groundwater Management Act requires that every basin determined by the state Department of Water Resources to be in overdraft, formulate a plan by 2022 and enact it by 2040. "Groundwater is being overdrawn by 15-20 percent in some places," said Richard Howitt, a consultant with ERA Economics and faculty member in the Center for Watershed Sciences at the University of California at Davis. "If we're going to stabilize that, we've somehow got to cut back on our net water use."

Cutting back on net water use means, in dry years, leaving fallow some cropland that currently produces every year, and allowing production (read: irrigation) only in wet years. Howitt expects lower-value crops like sugar beets and tomatoes to be affected first. Throughout the country, a lack of regional and institutional infrastructure for irrigation has thwarted reliable access to water.

In eastern Washington, Urban Eberhart grew up planting and irrigating the apple and pear trees on the family farm. He still remembers the first time he heard the word drought. It was 1977, and his father had taken him to a meeting about the water supply in the Yakima River Basin. Today, Eberhart manages his local irrigation district (one of over 100 farmer-run irrigation districts in the state) and grows 70-plus acres of apples and pears and about 20 acres of hay. Irrigation is essential due to the dryness of the climate there.

Washington's climate and water resource challenges are being watched by irrigation professionals nationwide because its watershed approach to managing water supply is among the most advanced in the nation. Until recently, however, infighting stymied progress.

The state that's known for its cherries and apples also hosts a wide and growing variety of produce. The Cascade mountain range divides Washington in half. In winter, the mountain range creates a rain shadow on the East side of the Cascades as precipitation moves east. Once it hits the mountains, that precipitation turns to snow, creating a dense snowpack that melts in the spring, runs into reservoirs, and is diverted into canals for irrigation on agricultural land.

The Yakima River, a tributary of the Columbia River (which comes out of Canada and loops around through WA and into OR), flows through a series of valleys in eastern Washington called the Yakima River Basin. The Basin encompasses about 6200 square miles and is home to 360,000 people. It supports a $4.48B crop, livestock, and food processing industry that includes a variety of tree fruits, hops, and wine grapes. Since 1977, there have been fourteen droughts in the Yakima Basin, which caused significant economic harm to produce growers. Scientists predict the Yakima Basin will experience 20 droughts in the next century. Eberhart and other stakeholders have spent the last 30 years trying to create a sustainable supply of water into the future with three main goals:

- To protect Yakima Basin fish and wildlife;
- To provide for Yakima Basin communities & industries; and
- To keep agriculture whole.

As Yakima Basin residents discovered in 2015, drought conditions aren't always caused by lack of precipitation. In the winter of 2015, the normal amount of precipitation fell on the mountains as rain. Instead of forming snowpack to refill reservoirs during the summer, the precipitation ran off through the river system into the ocean, which resulted in a drought. The "snowpack drought" caused severe damage. Streams dried up and left fish stranded. Farmers couldn't irrigate, which resulted in crop losses.

"When you have a shortage of water, and you have competing needs of fish and wildlife and agriculture, it results in arguments that land in court," said Eberhart. "We'd been fighting against each other in the Yakima Basin for decades. That wasn't working."

Enter the Yakima Basin Integrated Plan, which convened representatives from each stakeholder group, including produce growers, state agencies, federal agencies, local government, national environmental groups, and the Confederated Tribes and Bands of the Yakama Nation. Previously, these groups would only meet in court with lawyers present. "The court decisions produced winners and losers," said Washington Department of Ecology's Joye Redfield-Wilder. "Nobody got what they really needed."

After collaborating to select a list of projects that would provide enough water for all into the future, they built the trust necessary to move forward together. Today, it's common for a farmer in the Yakima Basin to speak up for restoring endangered salmon, and for a Yakima tribe member to endorse the merits of irrigation. The stakeholders are preparing to build water storage reservoirs to provide a reliable irrigation resource as well as lots of water for fish flow, and irrigation.

This involves raising the level of the Cle Elum Reservoir by three feet by putting extensions on the radial gates. Doing so enables the lake to hold an additional 14,600 acre-feet of water specifically to help with fish flows, which means that 14,600 acre-feet of water will not be taken away from agriculture. "We have some other places where we're talking about building a 160,000-acre reservoir-- half of it will go for fish flow, and half will go for agriculture, so 80,000 acre-feet of water," Eberhart explained. The working group is enhancing water conservation using several measures:

Additional lining, sealing, and piping of canals and laterals;

Building groundwater storage to capture the water that's flashing off in the wintertime;

Utilizing the existing canal system at a time of year it wasn't used historically, in order to divert winter and early spring runoff to groundwater infiltration areas where it will recharge the aquifer.

"We're coming up with a holistic approach that helps assure a secure water supply in the future for irrigating crops," Eberhart reported. "In addition to building storage, we're also doing habitat improvement... water conservation, working on fish passage, finding ways to consolidate farmers' diversions. Everybody still gets to irrigate, but we're finding ways for the fish to get where they need to go. When the dams were first put in, it wasn't a priority for the U.S. to put in fish passage and fish ladders. Now it is, so with our plan - the very first basin-wide integrated plan of its nature in the country - we are finding ways to protect agriculture and fish and wildlife at the same time."

In 2013, Washington state passed legislation supporting the Yakima project and committed to funding up to 50 percent of the project costs into the future. That amounts to $2 Billion (of the total of $4 Billion) over the next

30 years that will pave the way into the next century for Washington agriculture to be able to survive. Because the Yakima River Basin Integrated Plan is so unique, Eberhart and his colleagues find themselves talking to groups throughout the U.S. about how they put the plan together. "It is really collaborative, and it is going to be the way of the future to have Ag and Fish & Wildlife working together to benefit each other," said Eberhart.

An example of the collaborative nature of the Yakima Basin Integrated Plan is the cost-sharing aspect. Many of the project's costs come from the need to update irrigation infrastructure that was built in the early 1900s or before. That infrastructure - canals, laterals, pipelines - needs to be updated to be functional into the next 100 years. Often, the expenses far exceed what the crops raised in that area can pay for in a short period. Farmer-led Irrigation Districts partner with Fish & Wildlife funding sources to pay for canal piping and canal lining, which saves water, creates capacity, allows saved water to be leased to supplement stream flows, and carries water away from the river for groundwater storage. The partnership has fostered innovative solutions - like finding ways to use the infrastructure created for agriculture for other purposes while still meeting the irrigation needs of area farms.

"When the snowpack drought occurred, and we were starting to see tributaries dry up, and we were seeing stranded pools of threatened species, we actually put siphons in our canal system as temporary life-saving measures for these fish. We siphoned water out of the canal and put it into these tributaries by borrowing water out of a river system, looping it through, bypassing the river for a little bit, putting it into the tributary and then sending it back around," said Eberhart. "That was done because of the collaborative process we'd be planning for the future. We knew that's where we were heading in the long run; we just had to do it as an emergency measure. Now we're working on putting in more permanent structures."

Ultimately, agriculture benefits when the infrastructure improvements that need to occur are funded by environmental partners who also benefit by utilizing the capacity created with that funding.

At UC Davis, Medellin focuses on how to protect the ecosystem to increase productivity and water supply reliability. "We manipulate the natural systems to produce crops," he said. "Agriculture has evolved, and now we use more inputs to increase productivity. We are sometimes keeping away the native species and allowing invasive species to come, which creates some imbalance in the native system. These practices often degrade soils and put pressure on water resources. Productivity increases on one side but degrades on the other side. In some places, no one is keeping the "checkbook" to record how much water is used and replenished, and so they keep pumping and don't recharge or monitor how much needs to be returned to the groundwater."

When the water table is very low, issues such as loss of storage capacity, subsidence, damage to irrigation infrastructure and increased risk of flooding may occur. Medellin said it's in every farmer's interest to understand and support the coordinated use of surface and groundwater, called Conjunctive Water Use, on a regional level and to encourage community investing in physical and institutional infrastructure to allow groundwater recharge. Protecting agroecosystems can increase long-term agricultural productivity and water supply reliability. "It's like having a savings account and using it for a not-rainy day," said Medellin. "You use surface water when it is wet and groundwater when it is dry."

CWU @ Your Farm

Depending on the type of water rights you have, you can divert water from streams to flood the fields during the winter and recharge the aquifer and use that recharge later in the summer or in a drier year. Allowing intentional flooding of your fields at certain times of year can be beneficial. Tests with intentional flooding in California alfalfa fields showed no substantial loss of yields. It also worked in almonds. "Some places are suitable for this practice, and some places are not," Medellin said.

As cropland irrigation acreage continues to grow annually, the potential to reduce groundwater resources increases. So does the potential to impact both private wells and base flows into streams and rivers. Mark Dubin, Agricultural Tech Coordinator with University of Maryland Extension and the U.S. EPA Chesapeake Bay Program Office said more precise nutrient, soil moisture, and irrigation management are important to managing nutrient losses to the environment while managing our water resources. "The increased availability of management tools, real-time information, and operator education will likely be key elements to achieving these goals," said Dubin.

"Trying to do what you've always done or what your grandparents did no longer works. It's not effective. It's about being better stewards of our water and paying closer attention," said Ken Goodall, of Reinke Manufacturing Company. "Things like the adoption of center pivot irrigation increases the efficient use of water so [you] can use less while still growing a good crop."

Inge Biscorner, of The Toro Company's Micro-Irrigation Business, added, "Technology, like state of the art drip irrigation, can help not only precisely apply water and nutrients to crops to enhance yield and quality, but also reduce inputs and groundwater and surface water contamination. Bottom line: profits increase and sustainability improves. But this requires education and investment, both of which must be funded by both the public and private sector for the U.S. to remain competitive with the rest of the world, and for the U.S. to remain food secure."

PART VII

UNDER PRESSURE
PESTS, PATHOGENS, AND WEEDS

"Climate change isn't just happening in one region or another. It's a global thing,"

~ Lindsey du Toit, Washington State University plant pathologist

15.

PESTS & PATHOGENS ARE ADAPTING
What This Means for You

In the last ten years, buckwheat has taken off as a crop in central Washington's Columbia Basin, where over 10,000 acres grow, some for export to Japan. Baby leaf spinach is another new crop in the area, testament to the fact that the Columbia Basin is gaining repute as a good area to raise a wide range of crops. Farmers are drawn to the amount of land available, the irrigation available, the expertise among growers, and the presence of numerous seed companies that bring contracts to growers. Yet with opportunity comes risk-- in this case, new diseases that prey on the area's recently established crops. Researchers at Washington State University have identified new pathogens in Washington's buckwheat crops that were never reported in the U.S. before 2015.

Pathogens are living populations of organisms with individuals that do better in different environmental conditions. These populations are constantly shifting. Plant pathologists monitor shifts in populations over time to be aware of what's happening with pathogens in relation to environmental changes and in response to farming practices. This is especially important in the context of climate disruption, because insects, pests, and pathogens worldwide are adapting to the changing climate. Researchers at the U.S.

Department of Agriculture (USDA), and Cornell, Rutgers, and Washington State University are monitoring changes in environmental conditions (some caused by climate disruption) and changes in pest populations throughout the country. As reported in the 3rd National Climate Assessment:

Insects are directly affected by temperature and synchronize their development and reproduction with warm periods and are dormant during cold periods. Higher winter temperatures increase insect populations due to overwinter survival and, coupled with higher summer temperatures, increase reproductive rates and allow for multiple generations each year.

In the northeast, the growing season is lengthening, and Cornell's Mike Hoffmann said this increases the chance of more generations of pests during that warmer period, which could result in greater damage. "The life span varies by pest," he explained. "But adding 15-20 days to the growing season gives some pests that much more time to develop and damage the crop."

From 1990 to 2012, plant hardiness zones shifted a full zone north. "We can now grow canola in NY, which we could not in the 1990s, for example. That also means that some insect pests that used to not survive over winter in certain areas are now able to survive because of the warmer winters," said Hoffmann.

Marjorie Kaplan, Associate Director of the Rutgers Climate Institute, shared reports of southern pests like tomato pinworm and a northward expansion of beet armyworm in New Jersey fields. "Similarly, we have had reports that peaches have been developing ahead of schedule and in recent years, peach rusty spot and peach blossom blight have been showing up in higher levels, occurring early in the growing season," she said.

Two pests that affect sweet corn are thriving in the changing climate. The corn flea beetle used to be wiped out during most winters in the northeast by cold temperatures. Now it's much warmer, so it survives and is a pest almost every year. The corn earworm used to arrive late in the summer on storm fronts. Now, northeast growers and researchers find corn earworm early in the season, because it's able to overwinter in the area's warmer winter conditions.

As reported by Nystrom, Venette, Dieckhoff, Hoelmer, and Koch in the scientific journal Biological Control in January 2017, brown marmorated stink bug, a pest of many important crop plants native to Asia, has invaded the USA, including the north central states. Researchers investigated the efficacy of classic biological control agents, since the stink bug's natural native enemies are not present in areas where the stinkbug is invasive.

Another recent USDA project investigated the extremely invasive pathogen Ca. Liberibacter asiaticus and a presumed virus that causes citrus chlorotic dwarf in Turkey. According to USDA literature, the pathogen is transmitted by a whitefly already widespread in the U.S. and thus has very high potential for invasiveness. The research was designed to improve both scientific knowledge of invasive pathogens of citrus and provide new tests useful for their detection and management.

While neither of the above projects specifically targets climate disruption, Washington State University's Lindsey du Toit points out that projects like these are highly intertwined with climate disruption. "Plant pathogens for the most part are very strongly influenced by environmental conditions. Their ability to reproduce, to sporulate, to thrive, to spread are very strongly influenced and on a very narrow scale. A few degrees temperature shift can make a huge difference in conditions becoming more favorable or less favorable for particular types of pathogens. That's true for insects as well. Some of the pathogens I work with are vectored by insects and the influence of environmental conditions, particularly temperature, heat, relative humidity - all that influences things like duration of leaf wetness, amount of dew, amount of rainfall, amount of frost or lack of frost, how long the season goes into the fall or how early it starts in spring."

du Toit works with vegetable seed crops grown in the pacific Northwest, including spinach, the cabbage family, table beet, Swiss chard, onion, carrot, and coriander. "With changing climate you see a shift in which pathogens tend to be more problematic or less problematic," she said.

2013, 2014, and 2015 were some of the hottest summers on record in eastern Washington, which made conditions more conducive for opportunistic onion bulb infecting bacteria and fungi. In addition, warm spring temperatures starting earlier in the year allowed some pathogens to get active earlier. The onion seed crops du Toit grows are biennial, so they must go through a winter to change from an onion bulb crop to a flowering crop of seed. If spring is particularly warm, as it was in 2016, diseases like downy mildew become active earlier in the season. That creates a longer window of susceptibility. "We had the worst problem with downy mildew last year that I've seen in 16 years of working with onions in the Columbia Basin," du Toit reported.

Managing pests and pathogens has always required awareness of the fact that these are living organisms influenced by the environment in which they live. They're also living organisms that adjust and evolve, within their biological limits, to environmental pressures put on them. Said du Toit, "Climate change is just one set of pressures. As scientists, we try to figure out what are the limits of those boundaries-- developing an awareness of the whole system we're working in. Climate is one big factor behind that."

16.

HOW TO ADAPT TO ADAPTIVE PESTS & PATHOGENS

Many things influence the intensity of a pest or disease outbreak. Environment is one of them. Increases in temperature and changes in precipitation patterns are already sparking new conditions and affecting the incidence of pathogens and the geographic distribution of diseases. This will be exacerbated as climate change becomes more intense.

"We see shifts everywhere. You constantly hear about new diseases showing up that weren't in an area before. Some of that may not be climate change; it may be pathogens moving via global trade, coming in on a crate or something," said Lindsey duToit, Washington State University plant pathologist. "Just because a new disease shows up that hasn't been there before doesn't mean it's the result of climate change. But certainly, climate change can influence the degree to which the newly introduced pathogen takes hold and flourishes."

In the Northeast, climate disruption has caused a 72 percent rise in heavy precipitation events. "That's pretty tough on farming," said Cornell's Mike Hoffmann. Because the growing season is now longer, thanks to more frost-free days, it seems logical to plant some crops earlier. However this can be complicated by intermittent wet and cool periods in the spring that set the stage for pathogens like late blight to take hold and wipe out early tomato crops. This is one example of a pest that has been aggravated by changes in conditions.

"Farming's already risky," said Hoffmann, who advised that as the weather gets more unpredictable and more challenging and creates greater unpredictability in pest infestations, farmers become more diligent, engage in more careful observation, more scouting, and more precision application of pesticides as needed. Higher temperatures will affect the efficacy of some pesticides so that must also be monitored. "If you spray one day and it rains for the next three, obviously that's another complication to be aware of," he added.

Insects, pests, and pathogens are often tied to a specific phenological stage of the plant. Thus they attack flowers, shoots, fruits or leaves. According to Peter Oudemans, plant pathologist at Rutgers University, pests adapt to specific geographic regions, microclimates, and even plant genotypes. Oudemans reported more range expansion among pests that attack cranberries and blueberries. As an example, cranberry plants begin to flower based on temperature and photoperiod, whereas pathogens respond more to temperature. Therefore, as temperatures increase during the spring months, the fungi develop with flowers and are better able to infect. "We are definitely seeing higher rates of infection by certain diseases," he said.

Oudemans focuses on developing better ways to control disease of plants. With blueberries, he and his colleagues use a phenology-based management system and have developed a blueberry calculator to predict key developmental stages. "This is, of course, climate based and helps growers predict when certain events will occur," he reported.

While this and other tools exist to help produce growers adapt as pests and pathogens adapt to climate change, du Toit, Hoffmann, and Oudemans agree that the key component to farming success in a changing climate is footprints in the field. "[Climate] change is happening, so you can't just look at your calendar," said Oudemans. "You need to be out there watching what's going on and not be afraid to report new things. It is likely that we will see new pests. Brown marmorated stink bugs, boxwood blight, Spotted Wing Drosophila, Phytophthora ramorum, emerald ash borer are examples of invasive species that we have seen. With climate change, we could see range expansions of insects like the southern pine bark beetle more frequently. Most of these problems need to be addressed by cooperation among grower, foresters, government and university extension programs. Organisms follow the food supply not property lines or county/state boundaries, so we really need to communicate and work together."

"Growers need to be scouting, not just for things they know [tend to] show up, but also for new diseases that they haven't seen before," said du Toit. "Some of the best growers I work with are those that go to commodity meetings, not just in their area, but the national meetings, the national allium research conference or the international spinach conference. They learn about what's happening in other areas and become aware of diseases or pests that they might not have had to deal with, but might now start to see in their region because of the influence of climate change on which organisms can become more problematic."

Du Toit added some of the best growers she has worked with don't just scout their own fields; they contact her if they hear about a problem that happened in another area to determine whether it's something they should scout for in their area. "I really think that the growers who tend to be most successful are those that are aware of what could be coming down the road," she said.

Across the U.S., organizations and extension agencies offer predictive tools for pests, infestation, and timing. In the Northeast, a system of weather stations known as NEWA (Network for Environment and Weather Applications) are available for purchase. NEWA stations provide day-to-day forecasts, linked to three or four dozen models on pests, pathogens, and other weather-related concerns. NEWA provides relevant information to fruit and vegetable growers as the climate changes and risks from pests increase. (Find more information at http://newa.cornell.edu.) NEWA users reported that--as a direct result of using NEWA pest forecast models-- they save, on average, $19,500 per year in spray costs and prevent, on average, $264,000 per year in crop loss.

In Oudemans' opinion, the key to responding to climate disruption for agriculture is adaptability. He said, "If your farming practices are set in stone you will probably fail."

9 Tips for Adapting Faster Than Pests & Pathogens:
• Monitor your fields for pest infestations.
• Diversify - spread the risk.
• Make sure any and all interventions make economic and environmental sense.
• Reach out to university extension specialists to learn how to monitor, assess and identify pests and pathogens early. This will help you prevent populations from exploding and becoming difficult to manage.
• Improve soil health.
• Rotate to non-susceptible crops.
• If your state extension programs are not offering regional monitoring, encourage it.
• Find tools online at http://climatesmartfarming.org/tools.
• Invest in a predictive tool like NEWA. Visit http://newa.cornell.edu for info.

17.
LEARNING FROM WEEDS

Weeds cause 34 percent of losses in global crop production. As climate disruption intensifies, weed pressure is expected to increase. Higher temperatures and changes in precipitation patterns are already creating new conditions worldwide. While higher levels of CO_2 can boost growth and yields in some crops, rising atmospheric CO_2 can also make the widely used herbicide glyphosate less effective and actually boost weed growth, adding to the potential for increased competition between crops and weeds. Several weed species benefit more than crops from higher temperatures and CO_2 levels. These include Kudzu, a vine that attacks trees and shrubs; Johnson grass, which attacks numerous crops; Morning glory annual vines, which attack trees, shrubs, field and vegetable crops; and Velvetleaf, the bane of corn.

Your immediate concern is, of course, managing or eliminating weed pressure this season. You can also benefit from connecting with ag extension programs, researchers at land-grant universities and other resources to learn what to expect from weeds in the changing climate and how to adapt. The first step is to define the term weed. Today, farmers are growing naturalized and native species as crops. Dandelion greens and burdock are two examples of plants commonly known as weeds that are now cultivated commercially. Both crops command prices comparable to other health products and sell in health food stores like Whole Foods and in major grocery store chains like Hannaford and Stop and Shop. Cornell University weed ecologist Antonio DiTommaso says a weed is any plant that is out of place and competes with the cash crop for resources. Sweet corn that pops up in a field of broccoli could be considered a weed. Thus, my crop may be your weed and vice-versa.

Managing Weeds

According to a report released by the United Nations Food and Agriculture Organization (FAO), herbicides have become the tool of choice in intensive farming because the weed control effect of tillage has proven to be insufficient in the long-term. "The problem of tillage is that by creating a good seedbed for the seeds, it creates the same conditions for the weeds. While weed seeds are buried deeply with the mouldboard plough, the same plough brings to the surface the weed seeds that had been buried the season before… Weeds propagating through sprouts or roots can even be multiplied by tillage implements, which only cut and mix them with the soil, so that the number of potential weed plants is increased. Through soil carried with tillage implements from one field to another, the weed population is also spread throughout the entire farmland."

Tilling for weed control is not the ultimate answer. It is often not necessary to eradicate the weeds completely, but only to avoid the setting of seeds and competition with the crop. Leaving weeds in a crop at a stage where the crop can suppress them and where there is no damage or problem for the harvest can help with managing other pests, such as termites, or ants, which in the absence of weeds would damage the crop. According to DiTommaso, weeds can also protect and restore soil, and effectively provide "surgery" when areas are torn up, burned or otherwise altered.

Completely non-chemical weed control is possible, according to the FAO. As reported in earlier issues of this column, such practices are already successfully applied in commercial farming. Weed germination declines in soil that has not been tilled for several years. The superficial weed-seed bank depletes. If no new seeds are added, seeds still remaining in the soil will not germinate as they will not receive the light stimulus or soil temperature fluctuations needed for germination. This happens because crop leaves filter out specific wavelengths of light that "tell" the weed seeds in the soil it is not in their best interest to germinate. "It turns off a specific pigment, a phytochrome, from turning germination on, and indicates to the weed seeds:

If you germinate, you'll have to compete with this crop above you," explained DiTommaso.

In bare, pristine soil, that light is not intercepted and the entire visible wavelength goes right through. "That's why I encourage vegetable growers who want to know how to manage weeds, to minimize how much bare soil is there and limit the amount of time bare soil is there," DiTommaso said. "If you've planted and you know you have weed seeds there, try to get the soil covered as fast as possible by either the crop leaves or by a cover crop. The weed seeds can detect the difference in how much far infrared light is getting through. You need to know; we're not just saying grow cover crops to physically keep weeds out, but also to reduce their germination. That's important."

DiTommaso studies the impact of climate disruption, principally temperature and precipitation alterations, on the impact of weeds in both cropping systems and natural areas. The increased frequency of extreme weather events (heavy rainfall, early frost, warm spells) in the northeast interest the researcher because such events are difficult to defend against. "How do weeds compete with crops during extreme events?" he asked. "I'm looking at species that are a southern weed species, like Sorghum halepense (Johnsongrass) and Ipomoea hederacea (ivy-leaf morning glory). These are major weeds in the South and Mid-Atlantic crops. Our concern is, with warmer temperatures, they'll be able to establish in places like the northeast where they previously didn't do well. These species are more able to adapt to extremes than crops."

Johnsongrass is a perennial weed that's closely related to sorghum. In Northern US and Canada, it's adapted to the colder climate by becoming an annual plant. In the Southern region, it's still a perennial that survives by spreading rhizomes. In the North, it survives by seed. "This ability to adapt to a new environment and eventually proliferate in it is something we should be looking for," said DiTommaso. "I look at weeds as a genetic resource for us. Because usually, their genetics aren't very restricted. Many of our crops are the same hybrid and the same line, so if it's susceptible to disease, we lose everything. Weed species, even individuals within the same species can have very different genetics - like an apple tree. We have to graft if we want to get the same apple."

DiTommaso is not a breeder, but he has been encouraging his colleagues who are breeders to ask: what can we learn from these highly adaptable weeds that can help our crops? "If one looks at what makes weeds what they are, their ability to adapt, particularly in ag systems, to disturbance is impressive," he said. Traits like fast growth, ability to grow in soils of differing fertility, drought tolerance, and resistance to certain diseases and insect pests have been bred out of crops through over-domestication, as breeders focused mostly on yield. "These traits that make these plants so problematic might be some things that plant breeders should be thinking about. They're wild relatives of our crops. There's something to be said about having some weeds around still. They do contain some traits that hopefully we'd be able to incorporate in some of our crops."

PART VIII

CROPS
WHAT HORTICULTURISTS KNOW

It is neither possible nor desirable to include a report for every crop that is being researched with regard to climate disruption. As Andrew Paterson, Head of the Plant Genome Mapping Laboratory at University of Georgia pointed out, even subtle differences in climate can determine the ability to realize high productivity of a particular crop consistently. That fact is evident from the impact of year to year fluctuations on the productivity and quality of many crops.

The majority of U.S. Government-funded research into crop response to climate disruption is focused on "world-feeders" like pulses, grains, and legumes. How most horticultural crops, like fruits, nuts, and vegetables will respond to climate disruption remains to be seen. "Climate change will affect every region of the world and every crop differently," said UC Davis professor Samuel Sandoval-Solis. "This is a multi-dimensional problem."

18.

PREPARING SLOW-GROWTH CROPS FOR A CHANGING CLIMATE

Pests, pathogens, and irrigation are every grower's challenge. These problems aren't directly caused by climate disruption. However, as climate scientists and ag researchers stated in previous chapters, the erratic and extreme weather caused by climate disruption has begun to, and will increasingly exacerbate agriculture's typical problems. Forecasts suggest pests, pathogens, and availability of clean water for irrigation will test farmers with greater frequency and intensity until climate disruption is reversed. The clock is ticking on our opportunity to reverse climate disruption.

Despite this, the pecan industry appears surprisingly unaffected. "To date, severity or negative impact of weather related stress factors do not appear to significantly differ from the highly variable historical record," USDA-ARS research horticulturist Bruce Wood explained via email. "Pecan is a highly adaptable species and tools exist that will likely enable pecan horticulture to adapt should abnormal climate related stresses materialize."

According to Wood, even the rise in atmospheric carbon dioxide concentrations benefits pecan by improving key physiological and horticultural traits. Pecan is a native (indigenous) species in the U.S., managed in natural stands as well as in planted orchards across the southern tier of states. Huge variations in soil and climatic conditions, as well as numerous co-evolved insects and diseases necessitate careful management that differs between geographic regions. In comparison, commercial production of other tree-nut crops in the US is concentrated in a relatively small geographic area with much less variability in abiotic and biotic conditions.

Walnuts and Almonds are non-native species that came from and thrive in Mediterranean climates. In the U.S., California's central valley provides that climate, and hosts 1 million acres of almonds and 360,000 acres of walnuts. But is California's Mediterranean climate becoming more desertous? With that state's recent drought, almond and walnut growers have been challenged with irrigation and air quality issues. (Researchers are exploring the viability of Arizona, Idaho, and Nevada for commercial almond growing.) Walnut growers are challenged by bacterial blight, which climate disruption may exacerbate or mitigate, depending on how it affects spring rain patterns. Finally, global warming raises concerns about insufficient chilling as winter temperatures rise throughout the growing region. Deciduous trees require a certain amount of chilling hours each year in order to leaf out, bloom, and ultimately produce a crop. California's nut growing regions are experiencing fewer winter days where temperatures fall low enough to meet those chilling requirements.

Pecan does have its physical limitations, but according to USDA-ARS pecan breeder L.J. Grauke, the greatest risk to the advancement of the industry might be a myopic focus on pecan production through the lens of past practice. "The greatest benefit for the pecan industry might be achieved by tree size reduction through both improved rootstocks and scions, which will improve both nut production and tree management," said Grauke.

This achievement will likely necessitate incorporation of crop wild relatives in breeding, broad cooperation in the testing leading to selection, and development of improved methods. Creating a database to house information available to a diverse research community will facilitate cooperation. Acquiring funds to pursue development of those tools will require the support of the pecan industry, which Grauke said, in the U.S. is regionally fragmented and focused on marketing rather than crop development.

Molly Brown, Associate Research Professor at University of Maryland's Department of Geographical Sciences, wondered if current investment in nut research is sufficient and if it will culminate in time to help growers who are planting crops now that won't reach maturity for five or more years.

Many crop breeding programs are underfunded, but California's walnut growers have funded UC Davis' walnut research program annually and ensured its stability with a $2.6M endowment. Growers also help shape the program and test new material. "There's a lot of feedback," said UC Davis walnut breeder Chuck Leslie. "It's not just university research people or even extension people. The growers themselves are pretty heavily involved in the research."

It takes fifteen or more years to produce a new variety of walnut. Walnut trees take four to five years to begin producing a full crop. An established orchard has a lifespan of twenty to thirty years or longer. As global temperatures rise and we see the multitudinous effects of trapped greenhouse gasses in earth's atmosphere, this means the climate in which a grower plants their nut orchard may be very different from the climate the trees experience when they reach maturity.

"There's been enormous change in maximum temperatures in the last five years," said Brown. "Do we need to develop new varieties that are tolerant of much higher temps? Or maybe we need to migrate these crops to places like Spain and France, which are cooler? There are really only two choices: develop new varieties, which takes a long time, or move the crop. As forecasts for temperature in 2030 and 2050 change, so do growing zones."

Increased summer temperatures could affect quality and increase water requirements. These may be bigger problems than loss of chilling. Climate disruption may also impact insect pests. "Husk fly, for example, has been an increasing problem that has been appearing earlier in the season in recent years," Leslie reported. "We don't know the reason but temperature changes could be a contributing factor."

Yet based on the research that he and his colleagues are doing, Leslie is confident that California's walnut producers will be fine. "They've got some play in terms of what they can get away with in terms of loss of chilling hours. There are lower chill varieties coming along. In future years, maybe we'll be chill deficient; there are probably cultural practices that could mitigate that. [Researchers] in Chile have worked with spray applications that can mitigate dormancy to some extent. There's probably more work to do on that. I don't think it's going to drive walnut out of the state."

While some growers are feeling the effects of ongoing drought, through innovation, the almond industry as a whole is thriving in the face of climate disruption. "Faced with the drought, the almond industry has continued to improve," said 3rd generation grower and CA Almond Board member Kent Stenderup. "Since 1994, we've improved our efficiency 33 percent. We used 33 percent less water to grow a pound of almonds, through adopting technologies and cultural practices."

In fact, Chris Messer, Director of USDA's National Agricultural Statistics Service Pacific Region reported in July 2017 that the almond crop forecast was up 7.9 percent from the 2015–2016 crop production of 1.9 billion pounds.

19.

HOW CALIFORNIA ALMONDS ARE THRIVING IN THE DISRUPTED CLIMATE

Global warming caused by human activity intensified California's multi-year drought. Further, according to a report in the scientific journal Geophysical Research Letters, drying is projected to increase in coming decades, and as the earth's surface temperature continues to rise, natural climate variability will become less able to compensate for the drying effect of global warming. Essentially, even if wet conditions soon end the official drought in California, the state will experience widespread drought-like soil moisture conditions throughout the century. This highlights the critical need for agriculture professionals to have a long-term plan for drought resilience.

Current predictions are that California will receive roughly the same amount of precipitation in coming years, with less as snow. This is not good news, unless policymakers can find a way to adapt California's water infrastructure. As snowpack in the mountains melts, it gradually releases water that the current infrastructure captures and redistributes during the dry summers. This accounts for about 30 percent of California's water. "If snowpack becomes less reliable, then all Californians – both farmers and non-farmers – will need to adapt and invest in new infrastructure, additional conservation, new ways of storing water to capture more of the rainfall in the winter for use during the dry season," said Gabriele Ludwig, plant physiologist and Director of Sustainability and Environmental Affairs for the Almond Board of California.

One way to capture rainfall is to allow it to be stored as groundwater. In 2016, the Almond Board launched a project to assess whether and where almond orchards can be used for groundwater recharge when rivers are running high. The Central Valley could store a lot of water below ground if practical ways can be found to:

- Get available water onto ground suitable for recharge;
- Avoid hurting crops that are present; and
- Avoid hurting groundwater quality.

"We are also trying to understand the opportunities to recycle water from multiple sources, such as municipal wastewater," Ludwig explained.

"The current global warming trends will certainly influence and probably change historical weather patterns. This uncertainty dictates the need to maximize environmental adaptability in new cultivars," said Thomas Gradziel, an almond researcher at UC Davis. "Historically (and even currently in many regions of the world), almonds have been grown as a dryland crop because of their capacity to produce crops even without irrigation. Almond is thus highly adapted for growth and productivity with minimal water."

However, summer irrigation greatly increases yield. Gradziel's greatest concerns about the effects of climate change on the global nut industry are sufficient water quantity and quality for irrigation and the loss of winter chill hours.

Despite these concerns, the almond industry is currently thriving. Almond experts cite innovation as one of the reasons. "The almond community has invested more than $60 million since 1972 to build a foundation of research on environmental, production and other issues to continually evolve best practices with changing concerns," Ludwig said.

Increased extremes in the weather may mean growers will need to invest in new varieties that can withstand a greater range of weather conditions. The Almond Board has a variety of research projects underway to help growers.

Breeding Climate Resilient Almonds

"Anything that passes muster in the research program needs to be tested in field," explained Ludwig. Variety trials are done with a grower who is willing to work for many years with a very complicated orchard set up and with researchers from USDA-ARS or UC-Davis. It can sometimes take more than 10 years in the field for weaknesses and benefits in a new variety to show up. Thus breeding, takes a VERY LONG time in perennial cropping systems, even with marker assistance that shows researchers which genes are responsible for various traits.

One of the Almond Board's many collaborators, Gradziel uses traditional breeding methods to develop rootstocks and scions with traits designed to allow more dependable cropping under variable climates, improve water use efficiency, and lower the need for agrochemical inputs. Although Almonds need fewer chilling hours than other deciduous fruit and nut trees, and will therefore likely be affected more slowly by the decreases in chilling hours, winter chill is required to achieve good flower density and fruit set. Gradziel anticipates the loss of needed winter chill will affect orchard productivity in coming decades.

As climate disruption causes more intense weather extremes, increased pest and disease pressure, and limited availability of water, Gradziel expects Nonpareil clones with low bud-failure potential will be most successful for growers. "Nonpareil remains the cultivar to improve on," he explained. "It originated in the 1860's in California as a dryland variety and has since shown exceptional adaptability and productivity across a range of environments and cultural management styles."

UC Davis recently released the Kester variety, which exhibits many of Nonpareil's qualities of adaptability and productivity, is fully cross compatible with Nonpareil and is less vulnerable to frost damage.

While the California almond industry has become more efficient, researchers are preparing for the possibility that the crop may need to move to Idaho, Nevada, or elsewhere in the US. "In the long-term, I expect Kester would outperform many current cultivars in these regions since frost will be more important and Kester has improved resistance/avoidance to frost damage," Gradziel said.

Moving Forward in the Disrupted Climate

To be resilient in the face of climate change, farmers across California will need to overcome and adapt to its challenges. Continued investment in research will fuel the next round of innovation to ensure California's farmers can continue to grow healthy, nutritious food while improving water efficiency. While more work is needed to adapt water infrastructure for a future with less snowpack and dryer soil, Ludwig said farmers can do their part to increase water use efficiency. "83 percent of almond farmers schedule irrigation based on tree need and/or soil/weather conditions instead of watering on a predetermined schedule. In addition, almond farmers in particular have adopted efficient micro-irrigation systems at almost two times the rate of farmers statewide," she reported.

As a result of efforts like these, almond farmers have reduced the amount of water needed to grow a pound of almonds by 33 percent since 1994.

The challenges almond growers see are the same or similar for other California producers. For example, other crops that require significant chilling hours may not grow in California in coming decades.

The lesson for growers everywhere is clear: learn about the anticipated results of climate disruption in your area and assess how those forecasts mesh with the needs of your current crops. If necessary, be open to growing new crops or moving your operation.

20.

BREEDING FOR THE NEW CLIMATE

As a fruit, nut or vegetable grower in our changing climate, your biggest challenge will be growing with warmer temperatures. Impacts of climate disruption include shifting growing seasons, heat stress on some crops, reduced chilling hours for perennial tree crops, drought-like conditions in many areas, and increased water salinity in other regions. While the 3rd National Climate Assessment reports that changes in the growing season may have positive effects for some crops, it also states that reductions in the number of frost days can result in early bud-bursts or blooms, consequently damaging some perennial crops grown in the U.S. Changes in the growing season will affect the types of crops that can grow in any given location.

Marjorie Kaplan, Associate Director of the Rutgers Climate Institute, believes scientific research will help develop more climate-resilient crops and management techniques necessary for adapting to a changing climate over the long term. For example, Rutgers scientists are engaged in a variety of plant breeding and selection projects to safeguard cranberries and blueberries against climate disruption-related stresses. UC Davis researchers are using new tools to develop genetic markers of drought and disease resistance in grapes.

A case study released in October 2016 by UC Davis and Pacific Biosciences, a private genomics sequencing Corporation, reported that warmer temperatures attributed to climate disruption are already being recorded in many prime grape-growing regions of the world. Current wine grape breeding focuses on flavor, not hardiness. In California, where the value of grape crops varies widely and is heavily influenced by local climate, it is especially important that new varieties be able to thrive despite warming temperatures. Dario Cantu, a plant geneticist specializing in plant and microbial genomics in the UC Davis Department of Viticulture and Enology, believes advances in genomics, including single molecule real-time sequencing, could provide insights into better ways to breed wine grapes for resistance to disease and drought. "In a worsening climate, drought and heat stress will be particularly relevant for high-quality viticultural areas such as Napa and Sonoma," Cantu was quoted as saying in a statement released by UC Davis.

Cantu was lead researcher on a cabernet sauvignon genetic sequencing collaboration between UC Davis and Pacific Biosciences., Cantu said. "This new genomics approach will accelerate the development of new disease-resistant wine grape varieties that produce high-quality, flavorful grapes and are better suited to environmental changes."

Using Pacific Bioscience's new genome sequencing process, Cantu is developing the genetic markers necessary to combine important traits into new varieties. Cantu said, "What we grow are the worst genotypes for many reasons, but they are the best for flavor. They are susceptible to diseases and soil-borne pathogens. They are absolutely not drought tolerant. But we still grow them because they have the names we want to associate with our wines."

Although grapes grow everywhere in California, the genotypes are the same in Central Valley, Napa, and Sonoma. This lack of species diversity creates vulnerability in the wine industry, especially while the effects of climate disruption are becoming more severe. For optimal growth, wine grapes will require the same kinds of frequently updated, region-specific varieties used for most crop plants. Cantu is working on the entire vineyard ecosystem, sequencing the host, the plant, the beneficial microbes, pathogens, and all other organisms associated with the plant. "We are moving the grape genomics field forward at an unprecedented pace because we have the best tools in our hands," he reported in the case study.

Nick Vorsa, Director at Rutgers University's Marucci Blueberry and Cranberry Research and Extension Center, has been working with berries since the 1980s. Early in his career, Vorsa collected over 600 varieties of cranberry to provide a gene pool and has identified four accessions. Today, the primary focus of his program is Fruit Rot Resistance in cranberry and blueberry.

Fruit Rot, the result of a fungal infection in the plant, has grown more prevalent with the hotter climate. Vorsa reported that Fruit Rot is probably the biggest threat to the industry. Although the researcher has identified four Fruit Rot resistant varieties, he said they're not worth growing commercially, because they lack the productivity required for commercial sustainability. With funding from USDA, Vorsa has spent over 15 years trying to introgress the Fruit Rot resistant genes into productive backgrounds of blueberry and cranberry. New Jersey cranberry and blueberry growers are participating in the breeding programs, as well as in Rutgers' pathology program. "We think we've made pretty good genetic gain," said Vorsa. "What we're trying to do is pyramid the Fruit Rot resistance genes from different backgrounds into one or two varieties to enhance resistance even greater."

Now in the third breeding cycle since 1997, the Rutgers team is using molecular genomics analysis to speed up the process. As a result, the researchers are now testing selections that have higher Fruit Rot resistance, good yield and other desirable traits, like insect resistance. "We've identified resistance in Florida blueberries that we're trying to bring into the blueberries that grow here. The Florida species wouldn't survive here, even though our climate is warming. It's not warm enough," Vorsa said.

In addition to Fruit Rot, New Jersey blueberries are now getting fungal diseases that were once relegated to the southern growing areas like North Carolina and Florida. Researchers suspect that the warmer climate has changed the principle fungi. Wisconsin is increasingly becoming more susceptible to Fruit Rot pressure as well. In response, Vorsa has steered Rutgers' breeding program for blueberries to address these newly prevalent diseases.

If all goes well, Vorsa expects to release Fruit Rot resistant varieties of blueberry and cranberry in about eight years. "There's the compulsion to try to get these out now and release them, but I feel we need to test these thoroughly before growers make that commitment. If growers make the commitment to planting these varieties and we haven't fully tested them, it could be very bad for the growers."

At a time when agricultural research is more important than ever, Vorsa said many breeding programs have been terminated. He urged your support of research programs in breeding, pathology, and entomology. "To maintain the industry breeding programs are essential, particularly in the huge climate shifts, to develop varieties that are adapted to this new millennium," he said. "New races of diseases and insects will continually evolve. When I started in 1985, two insects were not even talked about, and now they're back. The notion that agriculture is static is a false one."

PART IX

CLIMATE SMART FARMING METHODS

If some of the solutions prescribed in this book seem contradictory to you, you're right. Every farm is different. Therefore, every solution will be unique. Certainly, similarities exist between your farm and your neighbors', or even between your farm and one on the other side of the world. Yet there may be major differences between your microclimate, your crops, your farming experience, and the history of your — and everyone else's — plot of land.

In the following pages, we look at how some climate-smart farming practices are being implemented around the world to great effect, and we consider how the post-carbon farm may look.

21.
POLLINATION CHALLENGES & STRATEGIES

Pollinators are responsible for nearly thirty percent of our nation's food supplies. Honeybees (Apis mellifera) support an estimated $15 billion in crop production, visiting fruits, nuts and vegetables including blueberries, cranberries, cucurbits, apples, almonds, onions, celery, beet, brassica and citrus. In California alone, more than 850,000 acres of almonds require more than 1.5 million honeybee colonies for pollination.

Yet populations of honeybees and other pollinators around the world are declining. In the U.S., approximately thirty percent of colonies are lost each winter due to the combined effects of pests, pathogens, environmental toxins, and poor nutrition. While climate change is not the sole cause of stress to honeybees, the links to climate disruption are clear. For example, the honeybee pest Varroa mite could produce many more generations in a warm year than one with seasonally cool temperatures. UC Davis entomologist Dr. Elina Niño said the added presence of the mites stresses honeybees. Niño also expressed concern that California's ongoing drought has reduced the amount of available forage for the honeybee population. "We're feeding our bees sugar and protein supplements. It's becoming more and more expensive to keep the colony going," she said.

For growers relying on pollination to set their crops, unpredictable weather patterns can lead to unpredictable pollination. honeybees are most active in warm, sunny conditions, explained Cesar Rodriguez-Saona of Rutgers Philip E. Marucci Blueberry and Cranberry Research Center. Intense rain patterns can reduce the foraging behavior of bees.

Zeke Goodband, Orchard Manager at Scott Farm in Dummerston, Vermont, brings in about one hive of honeybees per acre at the beginning of the blossom bloom. Yet climate change has disrupted the growing season, requiring Goodband to adjust his pollination practices. In recent years, unusually mild starts have caused the farm's 120 varieties of heirloom apple trees to break dormancy as much as a month ahead of historical patterns. Subsequent returns to "normal" cold or freezing temperatures damage apple buds and blossoms. In cold snaps, imported honeybees don't do the job. "The honeybees seem to have the work ethic of teenagers — start work later in the day, stop earlier and seem to need near perfect conditions to really work," Goodband said.

Return of the Native
Before the advent of trucking honeybees around the country in 1907, produce growers without their own hives relied on wild native bees to pollinate their crops. There are approximately 4,000 species of wild bees documented in the U.S., and today, estimates for wild bee pollination services are at $3 billion.

"Many of our crop plants depend on bees and wild pollinators for fruit set," wrote University of Florida entomologist Glenn Hall on his website The Bees of Florida. Native pollinators can pollinate almost any plant, but they prefer certain plants over others.

In our changing climate, more growers are returning to wild pollinators to ensure the success of their crop. In New Jersey, USDA's Natural Resources Conservation Service assists local farmers in designing, installing and maintaining native pollinator habitats as conservation practices. Funding for these habitats is available through the various financial assistance programs offered by USDA-NRCS.

In Vermont, Goodband is relying more on wild bees for pollination because they work in cooler temperatures and tolerate wind and wet weather. "Native bees just seem more resilient and able to deal with these changes," he reported.

In Dedham, Maine, wild blueberry grower Gail VanWart discovered the same thing. Long stretches of wet weather during spring pollination had become an increasing problem for VanWart. "Honeybees will not fly when it's rainy, windy or when the temperature drops below 50 degrees Fahrenheit," she explained. "Spring is when the bloom comes on the wild blueberries, and they have just a few weeks to be pollinated before the bloom falls off. Any blossom on the wild blueberry plant that does not get pollinated will not become a blueberry."

VanWart switched from honeybees to native pollinators, so she could count on pollination happening, even in varied and inclement conditions. "We gained insight on how nature intended crops to be pollinated," she shared. VanWart also turned her wild blueberry farm into a sanctuary for native pollinators to exhibit their worth in crop production.

Since making the switch, VanWart has seen increased pollination and thus increased yields. Financially, "going native" has been a boon. "Native wild bees do not cost us any time or money to keep, so it is far less costly than keeping honeybees, especially in a northern climate where they have a hard time overwintering," she reported.

Encouraging Natives

Anne L. Nielsen, Extension Specialist in Fruit Entomology at Rutgers University, manages flowering weeds between trees in apple and peach orchards to reduce the number of bees foraging in the orchard after bloom. "Ongoing research in my lab has been on managing ground cover to reduce pesticide exposure to foraging bees — both honeybees and natives," she explained.

Josh Campbell, a native pollinator researcher in Dr. Jamie Ellis' Honey Bee Research and Extension Laboratory at University of Florida suggested growers make their cropland as attractive and pollinator friendly as possible by:

• Following exact label instructions for pesticides to minimize impact on pollinators;

• Using a no-till practice (even outside of the crops) to enhance native bee nesting structure for the 70 percent of native bee species that are ground-nesters;

• Planting native wildflowers that bloom at different times to provide nectar/pollen throughout the season, and;

• Leaving field margins unmowed and unplowed, because many "weeds" are highly attractive to honeybees and native pollinators.

To enhance the biodiversity of their orchard ecosystem, Scott Farm planted a plethora of wildflowers and a diverse array of tree fruit, shrubs and vines. "With 120 varieties of heirloom apples we're pretty well spread out with bloom times and harvest," Goodband shared. "We think our growing philosophy that recognizes the orchard as a potentially wildly diverse ecosystem will help us 'weather' the changes in climate."

In Maine, VanWart makes sure to have native plants around to provide native pollinators a diversified diet from spring to fall. "It's easy," she said. "It is just letting nature do what it was supposed to."

22.
REGENERATIVE AGRICULTURE
An Opportunity For Course Correction

"Nature regenerates, if you disturb it then leave it alone. Our [traditional] practices degenerate."

~Andre Leu, President, Board of Directors, International Federation of Organic Agricultural Movements (IFOAM)

For more than forty years, Andre Leu has educated farmers on every arable continent. He's spent a lot of time teaching and learning from African farmers in Ethiopia, Kenya, Uganda, Tanzania and Namibia. His mission, it seems, is to inspire other farmers to take their power back and do what only they can to reverse climate disruption.

Stabilizing CO_2 in the atmosphere now is critical. According to National Oceanic & Atmospheric Association, it's higher than at any point in the last 800,000 years. As of July 2016, carbon dioxide levels were at 404.39 parts per million (ppm), a 3.08 ppm increase over July 2015. While CO_2 levels do fluctuate daily, the average annual increase since 2005 has been 2.11 ppm. If the greenhouse gas continues increasing at this rate, then levels will be at 430 ppm by 2030. The tipping point, where scientists say we will lose the opportunity to reverse climate disruption, is 450 ppm.

"What we need to do now is start at this stage to remove 15.52 gigatons [of carbon dioxide from the atmosphere] per year," Leu said in a speech delivered at the Northeast Organic Farming Association's summer conference on August 2016.

Leu advocates for the use of organic farming techniques to mitigate against and adapt our food systems to the effects of climate change. A fifth generation farmer, he now operates a 150-acre fruit farm in Australia with his family. "As farmers, our job more than anything is to maximize our production systems so we can maximize sun energy and turn it into life," Leu said. "What is really important here is that as we look at our solar voltaic cells. At this stage, we have no science that can collect energy as efficiently as photosynthesis. Quantum physics may help solve it, but it hasn't yet."

Leu said the answer is regenerative agriculture, or what IFOAM is calling, "Organic 3.0 Systems." According to the data Leu presented, by converting just one-fourth of the world's farmlands to regenerative agriculture, we could stabilize the world.

What does this mean exactly? There are many ways to engage in regenerative agriculture, but Leu panned some common assumptions about methodology. "People say 'no till is better than tillage.' That's a data-free assumption. The only paper I read that had data compared organic tillage with Round Up ready no-till. At every measurement, organic tillage beat the Round-Up ready no-till."

The key to regenerating the land, according to Leu, is to copy nature. This does not mean eliminating farm machinery or other modern comforts and tools. Copying nature in an agricultural system includes:
• Regenerative grazing,
• Plowing correctly,
• Using native plant species to attract and destroy pests and pathogens (Push-Pull Agriculture),
• Fertilizing correctly.

Grazing

Is grazing an essential part of regenerative organic agriculture? No, but all ecological systems have both plants and animals. Separating plants from animals via feedlots and other big agribusiness practices creates a total disconnect from the natural order. It's not an ecological system.

The more biodiversity you put in your system, the more output you can get. Leu shared an important equation:

water + CO_2 + sunlight + photosynthesis = glucose + oxygen

"You want to maximize our capture, maximize the use of that energy principal system," he explained. "When you put animals in, you're reusing parts now that you're not using without animals, so you produce more food. And you produce urine and manure, which you can put back into the soil."

The chemical process that happens in an animal's rumen is another weapon against climate disruption. The air they breathe goes into the gut, where microbes break down and synthesize carbon dioxide and nitrous oxide. What comes out the other end now is high nitrogen – i.e. fertilizer. Regenerative agriculture is about trying to understand and emulate natural cycles, so that systems can regenerate.

Even growers who prefer not to keep animals on the farm can engage in regenerative agriculture, by including the other three aspects: correct plowing, correct fertilizing and using push-pull methods. However, Leu believes the system works better with animals. "It's not too hard to put in a few chickens," he said. "Chickens will go through and eat the bugs, and be part of the ecology and you've got eggs every day and protein - costs you nothing."

Plowing

"There's this mythology that plowing is the thing that oxidizes the carbon and releases it into the air. But if we plow correctly, we can increase carbon sequestration," Leu shared at the NOFA conference. Leu said that synthetic nitrogen fertilizers are the main cause of carbon loss (Other scientists quoted in previous chapters disagree). For every pound of nitrogen placed in the soil, soil microorganisms need to eat 15-30 pounds of carbon, so it turns into carbon dioxide.

Through a form of tilling called deep ripping, the soil is opened at a deep enough level to allow carbon to get carbon deep into the soil. Carbon that's deep in the soil is most stable. We're also seeing organic no-till or low-till methods. The combination of deep tilling with no-till or low-till is effective.

Fertilizing

"If you have 9,000 pounds of nitrogen in your soil, why do you have to put synthetic nitrogen in it?" Leu asked. "At the beginning, it helps the plant grow big, but by the time the plant really needs it to set the fruit, the synthetic nitrogen is dissipated. In natural ecosystems, the majority of plants get their nitrogen from the earth. We don't need to go fertilize forests with urea or synthetic nitrogen."

Native plants & Push-Pull Agriculture

One thing Leu learned from the African farmers is the value of what they call Push-Pull agriculture. In this method, growers intercrop with specific varieties of grass and Ticktrefoil that are native or adapted to the region. According to Leu, Push-Pull is superior to cover cropping.

"I read about these things; you read lots of things," Leu said. "I want to ground proof them, see it with my own eyes, so I know what is really happening." When Leu saw what farmers in Kenya and Ethiopia were doing on their farms, he became a believer.

23.

PUSH PULL

How Peas & Grasses Fight Climate Disruption

Corn is one of the world's major food crops, and with more than 90 million acres of corn growing in almost every state in the nation, the U.S. is the world's largest producer and exporter. Like so many important crops, corn has its nemesis: the stem borer moth. This pest devastates corn in stages. First, stem borer larvae feed on the leaves of the maize plant, then they bore into the stem. Using pesticides to control stem borer is usually ineffective, as the chemicals cannot reach deep inside the plant stems where stem borer larvae reside. Another pest, the parasitic invasive species Striga (witchweed) attacks corn and other globally important agricultural crops, including sorghum, sugarcane and rice. Striga is native to Africa and Asia and was first identified in North Carolina around 1956. As with stem borer, use of chemical herbicides against Striga is ineffective.

These pests don't feed exclusively on U.S. crops. According to reports from International Centre of Insect Physiology & Ecology (ICIPE), preventing crop losses from stem borers and Striga weeds, and improving soil fertility in eastern Africa alone could increase cereal harvests enough to feed an additional 27 million people. Now, some U.S. producers are learning about the Push-Pull method developed in Kenya that prevents such losses and may help farmers everywhere to protect their crops in the erratic weather created by climate change.

Learning from Nature

In Africa, economic barriers to conventional pest resistance led researchers at ICIPE to work with farmers in Kenya and other partners to look for clues in nature. By examining how native grasses deal with pests and weeds, the researcher/farmer team discovered certain plants that can more effectively and less expensively eradicate certain pests.

In a process scientists call "selective allelopathy," some plants actually suppress others. Desmodium, commonly known as Ticktrefoil, not only suppresses weeds, but also conserves the soil, exudes anti-xenotic allomones to repel pests, and provides high protein stock feed. Desmodium is a member of the pea family with native roots on every arable continent. In North America, there are 76 varieties of the plant. Ticktrefoils are also useful as living mulch, as green manure, and as a climate disruption mitigator, as they improve soil fertility and reduce greenhouse gasses in the atmosphere via nitrogen fixation. (Most also make good animal fodder.)

Napier grass works well in its native sub-Saharan Africa as a trap for ht corn borer. The sharp silica hairs and sticky exudates on the Napier grass kill the stem borer larvae when they hatch, breaking the life cycle and reducing pest numbers.

With this information, ICIPE and its partners developed the "Push-Pull" growing method, which relies on Napier grass and Desmodium to protect crops. To date, push-pull has been adopted by over 131,229 smallholder farmers in East Africa where maize yields have increased over 300 percent with minimal inputs. The process and science behind it are described at push-pull.net:

"The technology involves intercropping maize with a repellent plant, such as Desmodium, and planting an attractive trap plant, such as Napier grass, as a border crop around this intercrop. Gravid stemborer females are repelled or deterred away from the target crop (push) by stimuli that mask host apparency while they are simultaneously attracted (pull) to the trap crop, leaving the target crop protected."

Desmodium produces root exudates; some of the exudates stimulate the germination of Striga seeds while other exudates inhibit their growth after germination. This combination reduces the Striga seed bank in the soil through efficient suicidal germination, even in the presence of graminaceous host plants. A perennial cover crop, Desmodium exerts its Striga control effect even when the host crop is out of season. Desmodium also conserves soil moisture, enhances arthropod abundance and diversity, and improves soil organic matter.

Desmodium and Napier grass protect fragile soils from erosion. These factors enable cereal-cropping systems to be more resilient and adaptable to climate change while providing essential environmental services and making farming systems more robust and sustainable.

According to Andre Leu, president of the board of directors of the International Foundation for Organic Agriculture Movements (IFOAM), farmers he met in Kenya and Ethiopia are successfully using push-pull technology on millet, sorghum, pulses, mango, tomato, lettuce, squash, potatoes and other crops. "In the middle of a drought, where the crops of 40 million people have failed, these farmers are thriving," Leu reported.

The Push-Pull System in action.
On the left, Napier Grass attracts (pulls) moths, so they lay their eggs in the grass rather than in the maize.

On the right, Desmodium repels (pushes) moths from the maize and suppresses weeds, like striga.

Photo provided by Andre Leu, edited by R.L. Fraser

Adapting Push-Pull for Your Farm
Conventional push-pull was developed in 1997 and introduced to a small group of Kenyan farmers in 1998. It uses Silverleaf desmodium (*Desmodium uncinatum*) and Napier grass (*Pennisetum purpureum*). In response to sub-Saharan Africa's increased drought pressures in recent years, ICIPE and its partners developed what they call "Climate-smart" push-pull in 2011 and introduced it to farmers in 2012. Climate-smart push-pull uses two drought-tolerant species: Greenleaf desmodium (*Desmodium intortum*) and Brachiaria grass (*Brachiaria cv mulato* II).

Sudan grass (*Sorghum sudanese*) is a viable alternative to Napier grass for U.S. growers that has been used successfully in push-pull agriculture. Desmodium and Napier Grass (*Pennisetum purpureum*) are native to Africa. But varieties of Desmodium and grass grow on every continent. Finding varieties that are adapted to your region may be a matter of trial and error or of seeking assistance from researchers at your local ag-extension office. Leu said intercropping with varieties that are native or adapted to your region is superior to cover cropping. "When we bring these systems in we find that locally adapted seeds are the best responsive for yield," he added.

Push-pull is already used in large-scale commercial systems. Leu, a native of Australia, where the typical farm is thousands of acres and the smallest paddock on his family farm was a square mile, claimed it's just as easy to put Napier grass and Desmodium around a one-acre circle of corn as it is to plant it around a thousand acres of corn if you've got the right machine and the right system. "It shouldn't be an either-or," he said. "It's matters of appropriate scale for who you are."

The simple methods involved in regenerative agriculture provide farmers a chance to correct the course our planet is currently on. "This is urgent," said Leu. "We have a once in a generation chance to stop this. We don't want to be the generation that failed our children and grandchildren. We don't have to invent any new technologies. We just have to scale up good practices. A new GMO costs $100M. If we took that $100M and used that for training, we could stop climate change."

24.

THE POST-CARBON CITY AND FARMING
A Chat With UCLA's Stephanie Pincetl

We need to decouple the idea that human well-being is predicated on endless growth. We need to consider what's produced, by whom, for whom and how. My vision is about addressing the fierce urgency of now, not about the smart city where tech growth fuels a city that does everything for us and puts us out of work. That won't address the issues. We need to rebuild cities that reflect our dependence and interact well with our environment. The post-carbon city is the key to planetary survival. If we shape it correctly, we transform ourselves and a future earth.

~ Stephanie Pincetl, **sp**eaking at UCLA's Earth 2050 Conference

I met farmer and professor Stephanie Pincetl at UCLA's Earth 2050 Conference in October 2016. Pincetl's lab collects and assesses data about urban ecology, as well as urban resource use (mainly electricity, gas, and water). Pincetl's presentation on the Post-Carbon City inspired me, so when we reconnected in summer 2017, I asked her to consider how a Post-Carbon Farm might operate. In our conversation below, she described to me what sounded more like a complete agricultural system.

RLF: Can you put your research on cities into context for an agricultural audience?
SP: We tend to forget that cities are inextricably interdependent with the hinterlands. Cities exist by virtue of things coming into them, whether it's concrete or apples. That's the big picture.
We are interested in how to look at those flows and begin to say, "these are the impacts on these hinterlands. These are the impacts of that resource flow. How can we think about mitigating those or reducing those or thinking about reuse and recycle, within what's already here? How do we better use water? How do we better use the organic waste stream?"
There are certain kinds of solutions for different problems. All places do not have the same kinds of challenges.

RLF: What does the Post-Carbon Farm look like?
SP: The post carbon farm looks like a lot more physical labor. It looks like reintroducing animals to farming. It looks like thinking about managing soils in a different way that is not using nitrogen fertilizers that are derived from petrochemicals. It looks like a much greater integration of the farm with the city, so that organic waste from the city can be recycled through compost and manuring for the farm. It looks like using solar energy more in food processing. It looks like preserving as much land around cities as possible for agriculture, in order to shorten the supply chain.

RLF: Tell me about your farm.
SP: My husband and I have a 20-acre orange grove in Ojai, ninety miles north of Los Angeles... in a small valley that has historically grown really superb citrus. It requires an enormous amount of irrigation and pumping of water that is done with electricity. It's an example of the kind of very complex trade-off that one should think about carefully in terms of the use of water and the use of energy relative to the process. If we don't grow oranges, where will the oranges come from? Those are the kinds of difficult questions that we're going to start looking at seriously.

RLF: Do you have thoughts about making your farm more like the post-carbon farm that could exist?
SP: We're doing things like mulching, which reduces evapotranspiration from the soil, makes irrigation less necessary. We don't make a living from this. We cover our water costs and the costs of farm management, because we don't manage it every day. We don't have the resources to make it look different. We would have to drill our own well. What are the implications for the groundwater resource? These transition questions are really hard to resolve on an individual basis. Part of the problem is to think about solutions as emanating from single individuals and adding them to a transformation. That is totally unrealistic.

A better question would be: What is the capacity of the Ojai valley growers to come together around the post-carbon transition? Not very high, because the alternative networks don't exist.

We've initiated other kinds of conversations, like the use of pesticides. That is a very difficult conversation, because you go against the very well-organized Farm Bureau which is in alliance with the petrochemical industry, the pesticide companies and the sprayers, so there's interest in maintaining the current regime even though it's not gonna work. Right? It's a power struggle, and in order to be successful you need a sea change among enough growers. The individual is relatively constrained in dealing with these kinds of questions.

RLF: Does the Post-Carbon Agricultural system change the location of farms?
SP: The distinction now between urban and rural is a little fuzzy. You have to consider what your food miles look like. If you're talking about the amber waves of grain, maybe you're talking about long-distance transport by rail that is electrified. If you're talking about produce, you can talk more about an urban food supply - greenhouses that are heated with geothermal, or kept warm with manuring and peri-urban intensive agriculture.

If you're talking about fruit crops, like stone fruit - some fruit grows in certain places, not everywhere. It's also about rediscovering crops that can grow in climates where they are more specialized. We have moved toward monoculture crops that are easier to grow anywhere, like certain species of apples. But there are other species of apples that may tolerate the cold better, or heat better.

It's about returning to the question of soil and climate and diversifying the species of plants that we use for food and expecting less high yields in some cases. Intensive Biodynamic farming has very, very high yields, and it's also very labor intensive. The response is regional. How Post-Carbon looks in different places will be different. Because climate and soils are different, growing conditions are different. We will get more appreciation for what that means when we eliminate or reduce the use of fossil energy, whether for fertilizer, fuel, or pesticides, herbicides, etc.

RLF: Is there room for hydroponic and soilless growing systems?
SP: If the whole life-cycle accounting takes place, the room for hydroponic growing will be very small. Where's your water going to come from? How is it going to get there? How are you going to filter stuff? All of that requires energy. If you can do it without fossil fuels, sure, but I don't think it's an answer. I think it's a technical innovation seeking a home.

RLF: What about fossil energy in the Post-Carbon farm?
SP: The timeframe is very important to think about. The transition is very important to think about. I think that there's a role for fossil energy in the world going forward. What it is, I'm not quite sure of yet. For example, plastics in medicine are pretty valuable-prosthetics, IV drips, and things like that. I think that going forward what we have to do is consider the highest and best use of fossil energy.

If there are farm applications in which fossil fuel can be used in an extremely effective manner, I'm not going say 'no way.' But I do think that we have become enormously sloppy with this very, very valuable resource. My feeling is that we should treat it like gold.

We don't waste gold. Gold is used for the right applications for gold. Most of the gold that's been mined in the world is still in circulation. So, if we think about fossil resource, of which there's a lot but becoming harder to extract and more expensive, and having all these very serious impacts. If we think of it as a valuable resource that has serious impacts, but that we can use parsimoniously and well, I think that's a good thing to do.

RLF: If there's been one theme to this conversation, I feel like it's that waiting for the one group or the one person to tell us the one answer is not going to work.

SP: There's not one solution. There's this drive in society to find a magic bullet to solve all the problems, but we live in places. The magic bullet is to reconnect with those places and act in partnership with those places.
People complain about "there are no seasons in Southern California." Excuse me - we have no seasons because we have occluded the changes that happen through over-irrigation. If we moved to a more native plant palette and got rid of the lawns, we would know the seasons. We would know it's summer it's really dry. The leaves are crinkled. It's deciduous in the summer. And we will know that it's spring because it's so lush and it's unbelievably green. We do have actual real seasons here, but we have distanced ourselves from our place.

25.

THE POST CARBON FARMING SYSTEM IN PRACTICE
a conversation with Jack Algiere of Stone Barns Center for Food & Agriculture

When I learned about Stone Barns Center for Food and Agriculture in 2017, I knew immediately that I had to visit. The 22,000-acre property was donated by the David Rockefeller family, who was pasturing cattle on it but knew the land had more potential. Stone Barns formed as a non-profit in 2003 with a vision for people to have contact with the food system. Jack Algiere came onboard to plan the operation and lead the team. The center, located in Tarrytown, New York, opened in 2004.

During my visit, Algiere showed me several micro-ecosystems that he and his team have cultivated for specific purposes on the farm. As you'll see, what Algiere and his team of farmers, apprentices and support staff are able to accomplish is bigger than what many farmers have the capacity to create. Yet, this system serves as a model of what post-carbon farming can look like. The practices employed here can be used by anyone with the skill and drive.

RLF: What's happening here?
JA: We have a productive farm. We're selling to the public, selling to restaurants, a very active production, very diversified system. Vegetables, grains, poultry, ruminants, pigs... every aspect of the work we do is all based

on the central body of agro-ecological principles. That has encouraged us to ask what are the indicators that suggest these principles are truly helping.

RLF: Do you consider Stone Barns a "post-carbon" farm?
JA: It's more important that we're showing a general balance of health in the system.
The big thing for us is that we participate in the data collection process as well as having the best practices that we think are encouraging stability, not just carbon. We don't want to keep putting inputs on; we know the problem that has created for us. So how can we get ourselves into a system where we're weaning ourselves off inputs as much as possible?
We're not trying to go back in time. We're trying to manage our resources. That to me is a way to get towards resilience.

RLF: What is a post-carbon agricultural system?
JA: In carbon, it's not a numbers game; it's about stability.
When you talk about climate issues, the soil is probably the one place where we can absorb atmospheric carbon. There's a huge debate whether we could capture 30 percent of what we put in. It may be a defeatist conversation. That said, every place can improve. Agriculture doesn't have to be the antithesis of conservation.
We can be doing things that are really beneficial to us and to the place over long periods of time. It's a balanced rotation of things, cover crops, etc.

RLF: Are you also breeding seed here?
JA: We have 500 crops in play - 1500/2000 different varieties of seed in our seed closet. We work with the most local and the largest international ethically minded seed companies out there— organic, a lot of Dutch, a lot of Japanese seed companies, NY, ME (Johnny's). Overlapping, recognizing what it means to have a crop that can grow well in this place and be resilient. We need to make it constant— it can be that prescribed that we have to have this condition or else. I grew up in New England. The reality is regardless of climate disruption, I don't remember a year in my whole life ever that was the same. It's not like the constancy of the central valley that has all the

synthetic infrastructure to make it so productive - in the long run, it's not sustainable.

Everybody got so used to this very replicable design. Now, it rains out of the ordinary, or there's drought. Being a New Englander, we never know what's going to come to us. We, as a people have a certain ability to adapt. We also have what you could say is marginal land, in terms of big ag taking all the "real" vegetable and meat production to the flat ground. We never had the space to get to the size that American ag really became. We still have that hint of what the old farm looked like. The diversified systems never disappeared in New England and NY.

People in Iowa may have never seen a small diversified farm in their whole life. It's a big paradigm shift. It seems so complicated to start talking about diversity when you only grow corn or soy for generations. It doesn't mean it can't be applied. Obviously, the Midwest is not a place for annual things; it's a perennial grassland, but we keep pushing against it.

This place may not be the best place to just raise cattle. It's actually better to be diversified, in the same way diversity encourages resilience. Better if your community knows you as a diversified farmer and is willing to shift the diet to what you have available.

RLF: You have a CSA right?
JA: We have 150 families in our CSA. We develop and expand agroecological farming practices. We also want to create a culture of eating that can support that. That is a key part of our work.

RLF: How do you manage your rotation of crops at Stone Barns?
JA: Rotation really is processional — one thing opening up to the next. The more we recognize what that looks like, the better.
This space is a seven-year rotation. It has a broad diversity of crops. There [are] actually two rotations happening, so fourteen different crops happening over the course of seven years. Some areas are carrots in the fall and lettuce in the spring, and it won't be carrots again for seven years. We started that area with a fallow clover, had an oat crop on it, just ran the chickens over it—

they eat the grass and any extra seed from the oats and clover, and then the clover grows up through and makes use of the manure that hits the ground. There is a sort of succession and a timing.

RLF: Where do cover crops come into play in the rotation?
JA: When you look at corn as a good example, when people aren't using cover crops — it's a scarcity model, [a feeling that] we don't have enough nutrients for another crop or a cover crop. Any other plants are competition, rather than generating better water infiltration, better nutrient availability, better aggregate stability, respiration — all of the things that actually keep us from having to add more nutrients.

Legumes provide nitrogen — turns atmospheric nitrogen into plant-available nitrate. Cover crops like rye & grasses, are biomass, so they're generally developing carbon, but they're also pulling and activating through the rhizosphere and the biome that lives under the soil all of those organisms are assimilating and making available lots of elements that are in the soil, locked up. This concept that people need to constantly add trace elements —we're in a glacial space. It's only 10,000 years since the ice retreated. There is more trace mineral in this ground than we'll ever know what to do with. But if you can't access it, and the only way to do that is through biology, then there's deficiency. So unless you turn these things on and keep them in a living body cycle rather than a liquid pour-through. It's the difference between looking at something as a static thing versus a living, moving, intentional thing.

We have a certain amount of material in the ground. With the exception of tender perennials like tomatoes, almost all vegetables came from weed species. They were the simplest things we could grow. They were the things that grew in the most disturbed environment. We just turned them into something productive and feel proud of that. They actually need disturbance to do well. Their job is a pioneer in that space, and then it's not — then stability takes place.

Especially in vegetables and the commodity annual — it's really exploitive, but its job in the whole of nature was actually just to seal the surface for the real healing to happen and the stability to take place in primary environments. These vegetables are not from primary environments. Parsnip, kale — they're all from when the ground gets scraped, those things come up. So, knowing who the plants are and knowing where we're going with this is part of knowing what they need to be healthy.

RLF: Are you talking about inputs now? Tilling?
JA: Compost. We make compost here on the farm. We manage all of our neighbor's waste materials as well — about 1,000 yards of animal manure, leaves from estates, woody biomass from the state park, no municipal materials. We compost and mix and produce particular blends for our [farm] and that's what we apply.

Compost by itself - people consider it fertilizer, which isn't correct. And it's not going to be soil. It's like fish food; it's energy. You add it to the space and it gets put into the fold. The things that live in the compost are not soil organisms, the soil organisms eat the composting organisms and they retain all the nutrition that was in those organisms, so it's like a bubble in a bubble. So when people set up gardens, it's very common to have bare ground and to spread a thick layer of compost and it's not correct. One of the most valuable things that happen with compost is the biology that's in it. It doesn't take a lot.

Europeans put as little compost as possible on the surface — and it's effective. It's necessary to use some compost, but a lot is pollution. The timing is also important. If you put it on bare ground, the biotic community is not there in the strength to use what you put on. Instead, a lot gets leached through, volatilized in the sun. The plants, like kohlrabi, have this root system that are activating what's in the soil. If the plant isn't there, there are other things going on in the soil.

If we put a cover crop on, we mow it off the ground, put the barley, then cut that, let the chickens eat it. It keeps it from leaching away. The timing of when we put the chickens on is important. They really are manuring what you feed them. That goes into the soil. It gets activated. It stays in there and grows. Then, we turn that clover in and we grow our crop in it. Same as with compost, it makes it much more valuable and stays there instead of flowing through and going to the ocean.

RLF: How does the composting process relate to climate disruption and climate resilience?
JA: Our role as the productive stabilizing force of agriculture is to produce off the space but don't just let it all flow away.
The more we learn about health, it's subtler. We have to keep evolving toward that subtlety — less addition. We dumped so much in because we just took those variables as constants. We'll always have enough water, we'll always have enough nitrogen, but now we know that's not true. One perspective is — it's a good corn year, so why worry.
Another is — I have a lot of work to do before I can get to self-sustainability, but I can work on co-creating community sustainability.
That's where climate change comes in. In the end, a lot of it has to do with water. Water is the thing that washes all the elements away.

RLF: So, you want to capture it and store it for later use...
JA: If you have terrible soils that don't have that stability and biology, etc., we lose all the fertility in the water — an entire summer of rain, an entire summer of total drought.
Resilience is that totally parched soil has had enough years of care and attention that it can still produce and hold some better water. And in a totally wet year, the soils can drain well enough to not totally sog up and be full of disease. That's resilience, not just solving for one problem. Climate disruption is so complicated in terms of what our expectations are going to be.

RLF: When I started writing about climate change (which we're now calling climate disruption) and its relationship to farming, I had no idea of the complexities of the subject. And communicating about it is so important.
JA: It's more than just understanding plants — it's understanding ecology, understanding what makes a cuisine, what a place is, and immediately starting a dialogue with your community: *What do you want to do with what we just gave you and how does that change what I do?*

Now we have a dialogue, and if we can sustain that over 25, 50, 500 years, then you have a cuisine. This is what happened. We don't recognize that. We don't have a corn cuisine even though that's our crop, because it really isn't sustainable even in the best possible practice. It's not the only option.

The clarity in that is we know, by itself corn can't sustain itself. It sustains itself because we continually add inputs. We're caught in this trap now of increasing additions and all kinds of things where we need to change the genetics all the time just to compete with the increasing pathogen pressures and weed pressures.

RLF: I have interviewed seed breeders who feel it's their job to feed the world, but they don't necessarily look at the impacts of their work on the ecosystem where they're planting their genetically engineered varieties.
JA: We're not trying to save the world. We're also not trying to feed the world or to look at [feeding the world] as the sole responsibility of a farm. We have a responsibility to just not ruin the place. Ag is supposed to be regenerative. We don't have a lot of inalienable rights as people. There are a few things though, like connection to the land. There are a group of people who feel that it's all just for us to use up. Some of us say it doesn't have to look like that. We can have that reverence and honor for place and it can still be generative for things around it.

RLF: I think a lot about whether ag can be regenerative *and* have reverence for animals. How would someone have livestock without eating them? Would it have to be a farm animal sanctuary where animals are foraging and grazing?
JA: Raising an animal for manure — what does that do for us from a vegan mindset?
Even when we get into the grassfed question, it always goes to: *Can we produce as much grassfed beef as we're producing in the CAFO?*
Why is that the first question? [We should be asking:] What can our ecology handle? What is a cow? What is it replacing? How does it fit into a design that we're working with?

I was a vegan for eight years but I also raise animals and I have a passion for that and grew up with that. The issue is around economics. To say that there is this insatiable demand for meat because people don't understand what its reason is. Or this concept that we need all that protein [to] keep us alive. But if we all stopped eating animals tomorrow, it wouldn't stop using animals... What is the purpose of the animal? And are we inclined to eat it? That's the second important question.

Look at vegetarianism. If we want milk — to produce it, you need a mother, which means you need a child, and if you don't want to eat them, what do you do with all those children? What is their purpose? The 40 billion broiler chickens that the world eats every year came out of this thing where we just wanted eggs.

We are so separated. What Stone Barns is [focusing on] is, in a sense, how do you get people grounded again? How do you get them in relation again? How can you understand and appreciate your food if you can't see where it's coming from? Kids come here and don't recognize a chicken. That separation has done everything — bad and good — for our society. We also took this naive approach to the simplicity of nature.

Agriculture, which was the boon of our civilization, is now arguably the thing that's going to take us down.

So we're doing it wrong — period. When we notice that at home, do we stop and choose something new, or do we just wallow in it, deny it or rationalize it?

RLF: We are now standing in this amazing, beautiful greenhouse with glass panels in the ceiling that open. And you designed it?

   JA: We've taken greenhousing to a different level... internal crop rotation soil design, year-round... We're using modern design, modern infrastructure and tech, with traditional growing techniques with modern varieties.

The glass ceiling panels in this greenhouse open to allow natural light and rain.

RLF: Varieties that you've developed. What's this tropical looking plant?
JA: That's turmeric; that's ginger. In the summer, we have crops that are low energy inputs - we don't want a lot of microgreens here in the summer. Ginger and turmeric are high-value crop but low touch. During the summer we're producing tons of veg outdoors, we don't want to compete with ourselves. In here, we want to produce a good complementary crop: baby peppers, lots of rows of dills and cilantros and things that are in constant rotation. We're in preparation now as new beds get together, as we finish crops, we're preparing new beds for spinach and carrots. It's all about timing, getting to a place of winter production.

This house has never been sprayed, organically or conventionally, because we're not chasing disease. We're dealing in health *all the time*. This is really important. Better plant health breeds better plant health.

[In here] there's very little data from minimally-controlled structures. You can have tons of control, but control leads to pathology because it's out of balance with the system, rather than working with it. Over time in a space like this, we've increased the organic matter from what was baseline at 4.5 percent to what has stabilized at 9 percent in 2009, and we've maintained that organic matter. It's my feeling higher organic matter does not mean better. It is not a game of money and numbers.

RLF: When do you apply compost?
JA: Apply compost to the cover crop and it eats it. When there's a live cover crop, we apply it to the top. We'll cut it once, all that material will lay on the ground; that makes it easier for us to spread compost over the top. The material falls down below the loose material on the surface. Volatilization is a concern. When compost is laid on the surface, any of the carbon, nitrogen, everything else just oxidizes and goes back into the atmosphere. You want it to stay cool and dark and you want it to enter. Once that turns in, in this case we might set up a bed. We pull out a live crop and leave all the roots in the ground and apply compost and kelp over the surface and broadfork it in. Most of it stays on the surface. It's not about rototilling or breaking. It's just about fracturing.

RLF: Disturbance but not disruption...
JA: It'd be easier to just pulverize but we want to get to a place where material goes in. Disruption and disturbance is entirely a part of this— the key is to only do it to a point that the plant needs. It takes experience to trust how little it actually takes. Often more inexperience leads to wanting to put more compost on. I can buy a product and it will save my thing.

Let's not touch it. How has consumerism fallen into this work? Rather than trusting engineering to solve our problems… there are engineering concepts in this house that have nothing to do with conventional or organic.

The building creates an atmosphere that we have control of. It has its limitations.

RLF: When do you open the roof?
JA: When it stops raining and is above 65 degrees the house will open automatically. There are hanging sensors in the middle at soil height. There's also rain gauges and weather stations. It's data collection. It's controlling a simple motor. All the watering is done by eye and by training. The glass ceiling panels in this greenhouse open to allow natural light and rain.

RLF: Why not use an automatic irrigation system?
JA: One of the things that I teach my managers and my perennial team especially is to never give up the mastery of the work that they do. They can try to make it easier for themselves, but if they're taking away from the mastery of the work, they're going in ten wrong direction. Putting a timer on the water doesn't enhance their understanding of the water, doesn't help in a disaster or a drought or whatever. Learning how to use a sharp tool, what it means to take care of your equipment, what it means to choose the right tool and the right time to use it. All these things are masteries. Why we're in a world where we think we have to be so simplified that anybody can just pick it up and run with it…

We're adapting our methods to what we're seeing. Our intention is to find a way out of what has essentially been not working.

A Farmer's Guide to Climate Disruption

RESOURCES

If you are considering adapting your farming practices with an eye toward climate resilience, the following resources may be helpful to you.

CA Healthy Soils Incentive

The California Department of Food and Agriculture Healthy Soils Initiative provides financial incentives for farmers and ranchers to increase soil carbon sequestration and reduce greenhouse gas emissions, with additional funding for on-farm demonstration projects that bring together agricultural partners, including growers, to field test management practices that provide climate benefits. Learn more at https://www.cdfa.ca.gov/oefi/healthysoils/IncentivesProgram.html

Ag & Climate Change Policies, Resources & Info

- Climate Hub http://www.climatehubs.oce.usda.gov
- Learn about NIFA's grant opportunities at https://nifa.usda.gov/apply-grant
- Preparing Smallholder Farm Families to Adapt to Climate Change pocket guides:
- Extension Practice for Agricultural Adaptation Managing Water Resources:
- https://agrilinks.org/library/pocket-guide-managing-water-resources
- Rutgers Climate Initiatives: http://climatechange.rutgers.edu/resources/climate-change-and-agriculture
- Each NIFA program releases a request for applications (RFA) with details on eligibility and how to submit proposals. See https://nifa.usda.gov/apply-grant.

- Read about NIFA's peer review process for grant applications and general writing tips for success at https://nifa.usda.gov/resource/nifa-peer-review-process-competitive-grant-applications.

- Watch NRCS demonstrate one important benefit of rotational grazing at https://www.youtube.com/watch?v=IqB4z7lGzsg

Land PKS

While the federal government no longer allows USDA scientists to discuss climate change or even use the terms "Climate Change" or "Global Warming," USDA has not yet removed news archives from the web. In 2017, USDA's Ag Research Service (ARS) announced the release of "Land-Potential Knowledge System" (LandPKS), a suite of smartphone apps that identifies and delivers information about specific soils. Jeff Herrick, a soil scientist with USDA in Las Cruces, New Mexico was part of an international team that developed, tested, and released the apps as part of a cooperative agreement with the U.S. Agency for International Development. LandPKS combines cloud computing, digital soil-mapping data, and GPS to provide information about the sustainable potential of land under current and future climate. "Herrick and his collaborators developed LandPKS because of the unique challenges that today's producers and land managers face in feeding a world population of seven billion people while also protecting soil, water, and other natural resources," the news release states.

A Note From The Author

Thank you for reading *A Farmer's Guide to Climate Disruption*. Please leave an honest review on the site where you purchased it. If you borrowed it from the library, please leave a review on Amazon.com.

ACKNOWLEDGEMENTS

This book would not be possible without the assistance, encouragement, or information provided by:

Ann Adams, Holistic Management International
Tom Akin, USDA-NRCS, MA
Jack Algiere, Stone Barns Center for Food and Agriculture
Inge Bisconer, The Toro Company
Kent Bradford, Distinguished Professor of Plant Sciences; Dir, Seed Biotech. Center, UC Davis
Jimmy Bramblett, USDA-NRCS, WI
Molly Brown, University of Maryland, Dept. of Geographical Sciences
Dayna Burtness, Nettle Valley Farm, MN
Josh Cambpell, Jamie Ellis Honey Bee Research and Extension Lab. Dept. of Entomology and Nematology, University of Florida
Rich Casale, District Conservationist, USDA Natural Resources Conservation Service, Santa Cruz, CA
Kenneth Cassman, Robert B. Daugherty Professor of Agronomy, University of Nebraska
Dave Chapman, Long Wind Farm, VT
Christine Clarke, USDA-NRCS, MA,
David Crafton, Six Oaks Farm, Norway, SC
Dr. Helen Dahlke, Dept. of Land Air & Water Resources, UC Davis
Kathleen Delate, Iowa State University
Antonio DiTommaso, Cornell Weed Ecology & Management Laboratory, Richard C. Call Director of Agricultural Sciences, Soil and Crop Sciences Section, School of Integrative Plant Science
Curtis Elke, USDA-NRCS Idaho
Kelly Flynn, USDA National Institute of Food and Agriculture
Michael Freeze
Bobby Galetta
Steve Gilman, NOFA Interstate Council
Matthew Gilbert, Department of Plant Sciences, UC Davis
Bridget Gnadt, Waukesha County Government

Ken Goodall, Reinke Manufacturing Company, Inc.
Zeke Goodband, Orchard Manager, Scott Farm, Dummerston VT
Thomas Gradziel, UC Davis
L. J. Grauke, Ph.D., USDA-ARS Pecan Breeding & Genetics, National Collection of Genetic Resources for Pecans and Hickories
Glenn Hall, Entomology & Nematology Department, University of Florida
James Hansen, climate scientist, Columbia University, formerly of NASA
Tom Harter, Robert M. Hagan Endowed Chair in Water Management & Policy, Dept. Land, Air, and Water Resources, UC Davis
Jerry Hatfield, USDA-ARS, co-author 3rd National Climate Assessment
Elizabeth Henderson, NOFA Interstate Council
Michael P. Hoffmann, Executive Director, Cornell Institute for Climate Smart Solutions
Cameron Holley, Associate Professor, UNSW Australia
Richard Howitt, Professor Emeritus in the Department of Agricultural and Resource Economics and faculty member in the Center for Watershed Sciences at the University of California at Davis.
Randy Johnson, PhD, National Leader, USDA Climate Hubs
Marjorie Kaplan, Associate Director of the Rutgers Climate Institute
Bruce A. Kimball (Retired Collaborator), U.S. Arid-Land Agricultural Research Center, USDA - ARS
Jack Kittredge, NOFA-MA, Many Hands Organic Farm
Rob Krause
Laura Lengnick, co-author 3rd National Climate Assessment, author of *Resilient Agriculture*
Charles Leslie, UC Davis Dept. of Plant Sciences
Andre Leu
Heiner Lieth, Plant Sciences Department, UC Davis
Brian Lockwood
Stephen Long
Bennett Lovett-Graff
Gabrielle Ludwig, Almond Board of CA
Vivian Marinelli, FEI

Josué Medellín-Azuara, Associate Research Engineer, Department of Civil and Environmental Engineering, Center for Watershed Sciences

Russell Morgan, USDA-NRCS, FL

Tyson Neukirch, The Farm School, Athol, MA

Anne L. Nielsen, Assistant Professor, Extension Specialist in Fruit Entomology, Rutgers

Dr. Elina L. Nino, Harry Laidlaw Jr. Honey Bee Research Facility, UC Davis

Dr. Chris Obrupta

Ana Otto, Arizona Farm Bureau

Andrew Paterson, Regents Professor and Head, Plant Genome Mapping Laboratory, University of Georgia

Stephanie Pincetl, UCLA

Renee Randall Willow Ridge Organic Farm, Wauzeka WI

Julie Rawson, Many Hands Organic Farm, Barre, MA

Chris Reberg-Horton, NCSU

Philip Robertson - W. K. Kellogg Biological Station and Department of Plant, Soil and Microbial Sciences, Michigan State University

Dr. Cesar Rodriguez-Saona, Associate Professor, Rutgers University, Philip E. Marucci Blueberry and Cranberry Research Center

Tom Rogers, almond grower, Madera, CA

Matthew Ryan, Cornell University

Dr. Samuel Sandoval-Solis, UC Davis Center for Watershed Sciences

Nicholas Saumweber USDA-NRCS, District Conservationist – Vineland Service Center, NJ

Erin Silva, Assistant Professor, Organic and Sustainable Cropping Systems Specialist, Department of Plant Pathology, University of Wisconsin-Madison

Pete Smith, FRSB, FRSE, Professor of Soils & Global Change, Institute of Biological and Environmental Sciences, School of Biological Sciences, University of Aberdeen, Scotland, UK

Kent Stenderup, almond grower & member Almond Board of CA

Carlos Suarez, USDA-NRCS, CA

Dr. Elwynn Taylor

Brise Tencer, Executive Director, Organic Farming Research Foundation

Richard VanVranken, Rutgers University
Gail VanWart, Peaked Mountain Farm
Nick Vorsa, Rutgers University
Vicki Westerhoff, Genesis Growers, Inc. St. Anne, IL
Bruce Wood, USDA-ARS, Supervisory Research Horticulturist, Southeastern F & Tn Research Lab, Byron, GA

I am also incredibly grateful to my Patreon subscribers, whose engagement provides both moral and financial support. Learn more at https://Patreon.com/RebekahLFraser.

BIBLIOGRAPHY

FOOD SECURITY AT YOUR FINGERTIPS
USDA website info re: Burke, John J Laboratory Director and Research Leader ARS, SPA

Potsdam Institute for Climate Impact Research press release (2015). What Would It Take To Limit Climate Change. https://www.pik-potsdam.de/news/press-releases/what-would-it-take-to-limit-climate-change-to-1-5degc

Hertsgaard, M. (2015). Climate Seer James Hansen Issues His Direst Forecast Yet. http://www.thedailybeast.com/articles/2015/07/20/climate-seer-james-hansen-issues-his-direst-forecast-yet.html

COMMUNICATING CLIMATE CHANGE
Jerkins and Ory. (2016). Outcomes from the 2015 National Organic Farmer Survey and Listening Sessions.
https://ofrf.org/sites/ofrf.org/files/staff/NORA_2016_final9_28.pdf

nifa.usda.gov

Merrill, J. (2015). CDFA Releases Healthy Soils Program Framework. http://calclimateag.org/cdfa-releases-healthy-soils-program-framework/

Greene, C. (2016). USDA ERS Organic Production Documentation. http://www.ers.usda.gov/data-products/organic-production/documentation.aspx

Peak Oil News. (2013). US Agricultural Sector In The US Uses Less Than 2% Of Total Energy Usage.

http://peakoil.com/consumption/us-agricultural-sector-in-the-us-uses-less-than-2-of-total-energy-usage

DISASTER AT YOUR DOOR: METHODS FOR CLIMATE RESILIENCE IN EXTREME TIMES

American Psychological Association. (2017). Disasters. http://www.apa.org/topics/disasters/

USDA National Agriculture Statistics Service. (2012). Census of Agriculture. https://www.agcensus.usda.gov/Publications/2012/Online_Resources/County_Profiles/South_Carolina/cp45075.pdf

Gallagher, J.J. (2017). Hurricane Harvey Wreaks Historic Devastation. http://abcnews.go.com/US/hurricane-harvey-wreaks-historic-devastation-numbers/story?id=49529063

USDA Farm Service Agency. (2015). NonInsured Crop Disaster Assistance Program. https://www.fsa.usda.gov/programs-and-services/disaster-assistance-program/noninsured-crop-disaster-assistance/index

SOIL: THE CLIMATE BELOW YOUR FEET

Kittredge, J. (2015). Soil Carbon Restoration: Can Biology Do The Job? http://www.nofamass.org/sites/default/files/2015_White_Paper_web.pdf

Ingham, E. (n.d.) The Living Soil: Bacteria. http://www.nrcs.usda.gov/wps/portal/nrcs/detailfull/soils/health/biology/?cid=nrcs142p2_053862

Kennedy, A. (n.d.) Bug Biography: Bacteria That Promote Plant Growth. http://www.nrcs.usda.gov/wps/portal/nrcs/detailfull/soils/health/biology/?cid=nrcs142p2_053862

UNFAO. (2015). International Year of Soils.
http://www.fao.org/soils-2015/en/

UNFAO. (2015). Soils Help To Combat Climate Change. http://saveoursoils.com/userfiles/downloads/1447750537-SOS_climate%20and%20soil.pdf

Veganic Agriculture Network. (2012). Veganic Fertility: Growing Plants From Plants.
http://www.goveganic.net/article205.html

Jones, C. (2014). Nitrogen: The Double-Edged Sword. http://www.amazingcarbon.com/PDF/JONES%20%27Nitrogen%27%20(21July14).pdf

Schonbeck, M. (2015). What is "Organic No-Till" and Is It Practical? http://articles.extension.org/pages/18526/what-is-organic-no-till-and-is-it-practical

BEYOND CARBON
Smith et al. (2008). Greenhouse Gas Mitigation in Agriculture.
from http://rstb.royalsocietypublishing.org/content/363/1492/789

Cassman et al. (2003). Meeting Cereal Demand While Protecting Natural Resources and Improving Environmental Quality.
https://www.annualreviews.org/doi/10.1146/annurev.energy.28.040202.122858

Galloway et al. (2003). The Nitrogen Cascade.
https://academic.oup.com/bioscience/article/53/4/341/250178

Ingham, E. (n.d.). The Living Soil: Bacteria.

http://www.nrcs.usda.gov/wps/portal/nrcs/detailfull/soils/health/biology/?cid=nrcs142p2_053862

US EPA. (n.d.) Overview of Greenhouse Gases.
https://www.epa.gov/ghgemissions/overview-greenhouse-gases#nitrous-oxide

WATER
Lengnick, Laura. *Resilient Agriculture*. British Columbia: New Society Publishers, 2015.

Pajaro Valley Water Management Association. (2014). PVWMA Basin Management Plan.
http://pvwater.org/about-pvwma/bmp-update.php

Lockwood et. al. (2015). Recycled Water As Part of A Larger Solution.
https://mail.google.com/mail/u/0/#inbox/154816f85739e7ce?projector=1

Richard E. Howitt, Duncan MacEwan, Josué Medellín-Azuara, Jay R. Lund, Daniel A. Sumner (2015). Economic Analysis of the 2015 Drought for California Agriculture.
https://watershed.ucdavis.edu/files/biblio/Final_Drought%20Report_08182015_Full_Report_WithAppendices.pdf

USGS. (2014). What is Groundwater?
http://pubs.usgs.gov/of/1993/ofr93-643/

Hatfield et. al. (2014). 3rd National Climate Assessment. http://nca2014.globalchange.gov/report/sectors/water

Ledbetter, K. (2016). AgriLife Research Eyes Vegetables For Improved Water Use.
http://southwestfarmpress.com/vegetables/agrilife-research-eyes-vegetables-improved-water-use?page=1

Chivian and Bernstein. *Sustaining Life: How Human Health Depends on Biodiversity*. Oxford University Press, 2008.

Hansen, M. (2014). Integrated Water Plan Moves Forward.
http://www.goodfruit.com/integrated-water-plan-moves-forward/

NWRA (n.d.) Washington State Water Resources Association. http://www.nwra.org/washington.html

Meek, R. (n.d.). Climate Change Impacts On Water For Horticulture.
http://www.ukcip.org.uk/wp-content/PDFs/EA_CCImpacts_Horticulture.pdf

Irrigation Associaton. (2017). Agriculture Exhibit.
http://www.irrigationmuseum.org/exhibit1.aspx?app=1

Bartels, M. (2017). The Waters Of The United States (WOTUS) Rule And Why It's Important.
http://www.audubon.org/news/the-waters-united-states-wotus-rule-what-it-and-why-its-important

USDA. (2017). National Drought Resilience Partnership End Of Year Report.
https://www.usda.gov/sites/default/files/documents/ndrp-january-2017-end-of-year-report.pdf

UNDER PRESSURE: PESTS, PATHOGENS, AND WEEDS
Hartung, J. (2017). Invasive Pathogens Of Citrus
https://www.ars.usda.gov/research/programs-projects/project/?accnNo=422986

Santacruz, E.N. et.al. (2017). Cold tolerance of Trissolcus japonicus and T. cultratus, potential biological control agents of Halyomorpha halys, the brown marmorated stink bug. *Biological Control*, 107, pp. 11-20.

Hatfield et. al. (2014). 3rd National Climate Assessment. http://nca2014.globalchange.gov/report/sectors/water

Bwalya, M. (2005). Conservation Agriculture: Does No-Till Farming Require More Herbicides?
http://www.fao.org/ag/ca/CA-Publications/Pesticide%20Outlook%202005.pdf

Cornell University Weed Ecology & Management Lab website: http://weedecology.css.cornell.edu/

CROPS: WHAT HORTICULTURISTS KNOW
Coppock, ed. *The Walnut Industry In California: Trends, Issues And Challenges.* University Of California, 1994.
http://aic.ucdavis.edu/publications/CAwalnuts.pdf

(2016). USDA NASS: Pacific Region Crop Production Report
https://www.usda.gov/nass/PUBS/TODAYRPT/crop0716.pdf

Valentine, K. (2015). How Much of California's Drought Was Caused By Climate Change? Scientists Now Have The Answer.
https://thinkprogress.org/how-much-of-californias-drought-was-caused-by-climate-change-scientists-now-have-the-answer-56ae9e33555f#.peh6bven8

Sahagun, L. (2015). California Drought: Climate Change Plays A Role, Study Says. But How Big?
http://www.latimes.com/science/la-sci-climate-change-drought-20150820-story.html

Williams, A. P., R. Seager, J. T. Abatzoglou, B. I. Cook, J. E. Smerdon, and E. R. Cook (2015), Contribution of anthropogenic warming to California drought during 2012–2014, Geophys. Res. Lett., 42, 6819–6828, doi: 10.1002/2015GL064924.

Polashock et. al. (2009). The North American Cranberry Fruit Rot Fungal Community: A Systematic Overview Using Morphological And Phylogenetic Affinities. *Plant Pathology.*
https://naldc.nal.usda.gov/download/36464/PDF

PacBio. (n.d.) Breeding A Hardier Cabernet: Smrt® Sequencing Provides Detailed View Of Grape Genome
https://www.pacb.com/wp-content/uploads/Case-Study-Plant-and-Animal-Sciences-SMRT-Sequencing-Provides-Detailed-View-of-Grape-Genome.pdf

CLIMATE SMART FARMING METHODS
Chris. (2007). The Bees Of Maine.
http://allthingsmaine.blogspot.com/2007/04/bees-of-maine.html

Swartz, B. (2015). The Maine Bumblebee Atlas Takes Flight. http://www.maine.gov/wordpress/insideifw/2015/04/24/the-maine-bumble-bee-atlas-takes-flight/

Desmodium. *Wikipedia.* Last modified August 25, 2018.
https://en.wikipedia.org/wiki/Desmodium

Sorghum x drummondii. *Wikipedia.* Last modified July 15, 2018.
https://en.wikipedia.org/wiki/Sorghum_×_drummondii

Pennisetum purpureum. *Wikipedia.* Last modified November 4, 2018.
https://en.wikipedia.org/wiki/Pennisetum_purpureum

USDA Plants Profile. Accessed 2016. http://plants.usda.gov/core/profile?symbol=desmo

USDA Invasive Species Info Center. "Witchweed". Accessed 2016. https://www.invasivespeciesinfo.gov/plants/witchweed.shtml

International Centre of Insect Physiology and Ecology. *The 'Push–Pull' Farming System: Climate-smart, sustainable agriculture for Africa.* Green Ink Ltd, 2015.
http://www.push-pull.net/planting_for_prosperity.pdf

Accessed 2016. https://www.co2.earth

Know Your Pasture Grasses: Brachiaria, A High-Quality Fodder Grass. *The Organic Farmer.* January 28, 2015.
  http://theorganicfarmer.org/Articles/know-your-pasture-grasses-brachiaria-high-quality-fodder-grass

"Push-Pull: A Novel Farming System For Ending Hunger & Poverty In Sub-Saharan Africa." Accessed September 2016. http://www.push-pull.net/3.shtml

ABOUT THE AUTHOR

Rebekah L. Fraser is a writer with a degree from Yale. Her articles, creative nonfiction, and personal essays have appeared in publications throughout North America. For ten years, she covered various aspects of agriculture and the food industry, including seed science, best business practices, produce marketing, and climate change vis-a-vis agriculture for *Growing Magazine: Produce for Profit*, *Blueprints*: a publication of the Produce Reporter Company, *Farming: The Journal of Northeast Agriculture*, and *Christian Science Monitor*. Fraser also writes fiction. *A Farmer's Guide to Climate Disruption* is her first nonfiction book.

To learn more about the author and her work, visit bit.ly/RLFReaders

Bee

www.ingramcontent.com/pod-product-compliance
Lightning Source LLC
Chambersburg PA
CBHW060642150426
42811CB00078B/2249/J